GRAVITY EXPLAINED

The Quantum Model of Motion
and
The Energy Cycle

Martin O. Cook

Editing and Contributions by
David Miller, Stephen Gibbons, Jeffery Davis, Jack Hylton Jr., and Bryan McPherson

ISBN-13: 978-1484025017
ISBN-10: 1484025016

Contents

Introduction

Gravity remains a mystery not because of what we have yet to discover, but rather, what we have failed to recognize from the clues we already have.

I understand why you would expect a book explaining quantum gravity to come from a great scientist like Albert Einstein or Stephen Hawking, or even a collection of scientists like the thousands who worked on solving the Higgs boson puzzle. Many physicists have tried and many physicists are still trying to piece together the gravity puzzle, but the fact remains that even to this day in our age of scientific and technological enlightenment, quantum gravity still remains a mystery.

How important is gravity? Gravity is the unifying mechanism upon which all of physics and life have evolved. It is the central influence in the creation of all things. Where would we be without gravity? So how is it that we have known about gravity for hundreds of years, (since Newton), known of its crucial importance, and still fail to explain how it really works?

The quagmire modern physics is presently trudging through in its futile attempts to explain the origins of gravity is directly linked to three obstacles that stand in the way: Newton's mistake, the misuse of time, and the lack of the big picture. These obstacles have created an impenetrable barrier that continues to stifle all who try to solve the gravity puzzle. Einstein's scratched out equations written with his dying hands testify that mathematics alone cannot solve this great mystery. What is needed is a theory like plate tectonics that pulls all the individual pieces of a puzzle together to form a more complete picture of how something really works.

Welcome to Quantum Gravity Explained.

Chapter 1

The Gravity Puzzle

How is it that scientists can tell us about the big bang, black holes, and subatomic particles, but fail to explain how the observable force of gravity really works? It has been over three hundred years since Newton pinpointed the force keeping order in the universe and apples on the planet, and after all these years, we still can't explain how gravity operates.

Many moons ago, when the realization of the preceding paragraph impressed itself upon my mind, I was dumbfounded. "You've got to be kidding!" I thought. How in the world is it that we still can't explain something as obvious as gravity. Einstein's general relativity did a better job than Newton's equations in predicting the results of gravity and the impact of gravity on light, but what in the world is space-time? Did Einstein make-up space-time to validate his equations? It sounds as mysterious as gravity. And in the end, space-time did not translate at all into quantum physics. And what about quantum physics? Quantum physics explains in detail the composition of an atom but completely fails to explain how that atom moves through space. Inertia? What is the science of inertia? Again, we are left with some mysterious action, like space-time, to explain the visible results of both motion and gravity. Ironically, the same theory that will explain motion will also explain gravity. We just have to hurdle three obstacles in order to introduce that theory.

The three obstacles standing in the way of solving the gravity puzzle are Newton's mistake, the misuse of time, and finding the big picture. Because we have failed to expose all three of these obstacles, quantum gravity still remains a mystery. The first obstacle is Newton's mistake. What was Newton's mistake and why has it had such a crucial impact on preventing a viable explanation on how gravity really works? This mistake is the most crucial obstacle standing in the way of explaining quantum gravity. It is because of Newton's mistake that science presently lacks a viable theory that explains how mass moves through space, the science of motion. The bulk of this book deals with hurdling this first obstacle. The second obstacle is the misconception of time. What is time, and how is a misunderstanding of it an obstacle to explaining quantum gravity? The third obstacle is the lack of a big picture. Try explaining the formation of the Rocky Mountains or the volcanoes of Hawaii without the big picture of plate tectonics. Or try explaining the reason we get rain without the big picture of the water cycle. Gravity is no different. Quantum gravity cannot be explained without the big picture of the Energy Cycle.

All three of these obstacles require a paradigm shift from our present perceptions of motion, time, and the flow of energy. And as we have learned from past experience, paradigm shifts do not come easily. For example, most doctors initially ignored the strict hand washing policy implemented by Ignaz Semmelweis in the 1850's. If they couldn't see it, they wouldn't believe it. Their mind's eye blinded them to realities existing outside the limitations of their conditioned paradigms. And Galileo spent the last nine years of his life under house arrest for scientifically supporting Copernicus' shift from a geocentric perspective to a heliocentric perspective. A sentence well deserved, right? When new ideas directly conflict with old, established ideas, the old, established ideas usually win out for a time.

This book is about solving the gravity puzzle, a puzzle with many individual pieces that have yet been fitted together to make a coherent whole. That is…until now.

Part I

The First Obstacle:

Newton’s Mistake

Chapter 2

Newton's Mistake

How or why do objects move through space? The colossal failure to answer this question of how objects move through space in the first place is the biggest blunder in physics. It started with Isaac Newton, was missed by Einstein and Heisenberg, and is still overlooked to this day. It is the main reason we have yet to explain how gravity really works.

Isn't it ironic that the very person who coined gravity, who probably did more for physics than any one individual except for maybe Albert Einstein, impeded its quantum explanation for centuries? What was Newton's Mistake? Newton didn't recognize that all objects are always in motion, even when they appear to be at rest. This is why he could never explain how gravity really works. He viewed objects, whether in motion or at rest, from their whole perspective and not from their individual parts. In doing so, he cemented the precedence that mass just mysteriously moves through space. This mindset prevented him from hypothesizing a theory to try to explain the science of motion to coincide with his laws of motion. Such a theory would have attempted to explain how objects move through space in the first place, potentially exposing the motion myth at the dawn of the scientific age. Not recognizing that all object are always in motion is the same mistake made by Einstein when putting forth his relativity theories. Einstein's theories do well to predict results, but they fail to incorporate quantum physics. The shortcomings of Einstein's theories will be addressed later. Physics still lacks a viable theory to explain from a quantum perspective how and why mass moves through space in the first place. We can no longer take for granted the inertia of mass.

So why blame Newton? If Galileo before him and Einstein after him are guilty of making the same mistake, then why refer to this colossal failure as *Newton's Mistake*? The thesis of this book is that this mistake, (whether made by Galileo, Newton, Einstein, or physicists today), is so crucial that it plagues our ability to explain how gravity really works. So a better question may be: Why not blame Newton? Newton authored the laws of motion. If the laws of motion are incomplete, if they don't tell the whole story about motion, should we not go back to the source? Newton's mistake perpetuated the motion myth and continues to impede our ability to understand how gravity really works.

Newton said he could see further than others because he stood on the shoulders of giants. What about his shoulders? Who is standing on Newton's shoulders? Essentially everyone who came after him. That is why his mistake has had a blinding effect for centuries. His mistake carried over to all those who stood on his shoulders. As a result, quantum gravity still remains a mystery to this day.

Why is exposing Newton's mistake so crucial to understanding how gravity really works? Simply put, gravity cannot be reconciled with quantum physics without a quantum model of motion to tie them together. A quantum model of motion explains how mass moves through space in the first place. From there, one can easily surmise how gravity really works. Since a quantum model of motion didn't cross Newton's mind, it somehow also slipped past the minds of all those who stood on his shoulders. By exposing Newton's mistake, we get to the source of the problem quicker.

Even though Newton's mistake has carried over to our day, I have no problem giving everyone a pass up to the inception of quantum mechanics. Even Einstein deserves a pass since he developed his relativity theories just prior to the formation of quantum mechanics. (He did spend the declining years of his life, after the inception of quantum physics, searching for a unification theory, but to no avail.) Current scientists face a different dilemma. Quantum physics has been around for a while. We have insights that weren't available to pre-quantum scientists. So why aren't we any closer to unifying gravity and quantum physics? The reason is because we are still stuck in the quagmire of Newton's mistake. The only way gravity and quantum physics will ever be reconciled is with a quantum model of motion to tie them together. A quantum model of motion will unite the differences between Newton's laws, Einstein's relativities, and quantum physics into a single, viable quantum theory. From the inside out, all three of these operate from the same quantum laws. A quantum model of motion will explain the role of quantum physics in the motion of objects through space, including the acceleration we call gravity. In order to explain how a quantum model of motion works, we first have to untangle the knot caused by Newton's mistake.

Newton's mistake ushered in a crucial misconception. The very person credited with discovering gravity hindered its ultimate explanation when he declared that a body at rest tends to stay at rest. For Newton, bodies were either in motion or at rest, which is understandable from his earthbound perspective. But from a quantum perspective, mass is never at rest; all mass is always moving though space. Even objects that appear to be at rest are still moving through space. What is important to realize is that it is the individual atoms that make up mass that are responsible for this perpetual movement through space. (This will be explained in detail in part three of this book.) This perpetual motion of all mass is *the great secret of physics* that links classical physics to quantum mechanics and is the key to understanding how gravity really works.

Think of astronauts experiencing a continual free fall as the space shuttle orbits the earth. As the astronauts appear to be floating within the confines of the shuttle, one astronaut passes an orange to another astronaut. As the orange makes the trek through space, its motion is visible by both astronauts. When the astronaut receiving the gift grabs it and places it in space next to him, the orange now appears to have stopped moving as it takes a position of rest next to the astronaut who placed it there. Is the orange really at rest? What appears to be at rest to the astronaut who just placed it in space next to him is really freefalling around the earth at 17,500 miles per hour. The orange is still in motion through space. Although objects can appear to be at rest, it doesn't mean they are not moving through space. They still have momentum. This is what Newton failed to grasp when he saw objects at rest on the earth's surface and categorized them differently than objects in motion.

Galileo also failed to visualize that all objects have momentum. When confined in a windowless cabin of a uniformly moving ship, Galileo said he wouldn't be able to tell the ship's speed relative to the land by applying what he knew about the laws of physics. He rightly reasoned that the laws of physics are the same for all inertial frames. All objects that appear to obtain a state of rest within any inertial frame mysteriously abide by the same laws of physics. In any inertial frame, if you throw a ball straight up, it will come straight down. If you put the same ball in a different inertial frame and throw it straight up, it will again come straight down. What he saw with his eyes was correct, but what he didn't see tells the whole story. Both Newton and Galileo were unaware that objects that appear to be at rest are never truly at rest. They are always in motion. As I will detail later, the collection atoms making up any mass are the source and cause of the momentum of that mass through space. So due to their limited perspective, Newton and Galileo were completely oblivious to quantum mechanisms operating within mass that account for the perpetual momentum of all mass through space.

What makes the perpetual momentum of all mass so indiscernible is that bodies at rest on the earth's surface really appear to be at rest. Their motionless bodies could not possibly be in motion and maintain that motion until an uneven force acts upon them. This is the vantage point from which Newton saw objects at rest. Like Newton, our minds do not naturally perceive that objects that appear to be at rest are actually in motion through space and the cause of that motion being the individual atoms that make up that object. The motion of objects that appear to be resting on the earth is tied to their gravity-bound status to the earth, sharing the same motion as the earth, similar to a comb sitting on a seat of a moving car sharing the same motion as the moving car. Once you slam on the breaks, it becomes apparent that comb's motion is independent of the motion of the seat upon which it appeared to be resting. This misperception that objects can obtain a state of rest has been passed down to us and continues to veil the mystery of gravity.

In reality, you cannot separate the physics of bodies in motion from the physics of bodies that appear to be at rest. All objects are always in motion even when they appear to be at rest. To emphasize this point, imagine a switch that turned off the gravity that accelerates objects into the earth's surface. With the effects of gravity shut off, the slightest nudge to any of these objects would cause these objects to move away from the earth's surface. The nudge wouldn't begin the motion; it would only initiate a slight variance of the motion they already exhibited when they were moving through space attached to the hip of the earth by the effects of gravity. As these nudged objects, once bound to the earth's surface, continued in their new momentum path, moving away from the earth's surface, one would witness that these objects always had motion. Their perpetual motion was cloaked by their gravity-bound status to the earth's surface.

Newton's mistake was his failure to perceive that all bodies or objects are always in perpetual motion through space and are never at rest. Within this perpetual motion, they can either accelerate or decelerate but at no time do they ever stop being in motion through space. Newton could have declared that all bodies are always in motion and are never at rest even when they appear to be at rest, and that all objects tend to maintain the same motion unless acted upon by an uneven force. This would have corrected a great misperception that has stifled a quantum explanation of gravity up to this point in human history. Perpetual momentum is the natural state of all objects and acceleration and deceleration are only temporary alterations to the equilibrium of perpetual momentum. The reality that all objects are always in motion is the *great secret of physics* that was firmly veiled by Newton's mistake.

Had Newton tried to devise a theory for the science of motion, he might have recognized his own mistake or at least planted the seeds that would inspire those standing on his shoulders to ask: How do objects move through space in the first place? In other words, what is happening on a quantum level that allows a body in motion to stay in motion? Or even more basic, what is happening on a quantum level that allows a body, any body, to be in motion at all or have any movement whatsoever? In short, what is the role of quantum physics in the momentum and movement of mass through space? By failing to hypothesize why mass moves through space in the first place, a faulty paradigm was adopted into the scientific community which implies that mass, which is made up of atoms, just mysteriously moves through space. Newton's mistake perpetuated the motion myth.

Chapter 3

The Motion Myth

Newton's Mistake.

Cemented...

The Motion Myth.

Which led to...

Einstein's Real Blunder.

Which led to...

Quantum Physics being Misguided.

Which led to...

The Standard Model remaining Incomplete.

Which leads to...

Quantum Gravity Still Unexplained.

During a crucial part of a game, a coach will call a timeout to assess the situation and prepare his players to execute at the highest level, desiring the greatest outcome possible. Presently, in theoretical physics, a timeout is needed to assess the direction in which it is headed. Even with the discovery of the Higgs boson, the Standard Model cannot explain how gravity really works.

It is with deep respect that I boldly point out a terrible oversight of all the great scientific minds preceding this work. Their failure to grasp that all objects are always in motion, even when they appear to be at rest, continues to erect an impenetrable barrier that prevents modern physicists from discovering how gravity really works. In order to rectify hundreds of years of tradition, I focus on Newton's mistake and how it carried over to Einstein, Heisenberg and the rest of the scientific community. They all failed to see this simple truth.

The motion myth is the idea that objects somehow mysteriously move through space without any explanation other than they just move through space. External forces can speed objects up or slow them down, but why or how? The motion myth illustration gives a brief overview of how modern physics is built upon the motion myth paradigm, from Galileo to the Standard Model. Any theory based upon the motion myth paradigm will eventually lead to a dead end. The quantum model of motion theory corrects the motion myth and is as important to physics as plate tectonics is to geology. The quantum model of motion illustrates how all objects have momentum, even when they appear to be at rest, and is the keystone for understanding how gravity really works.

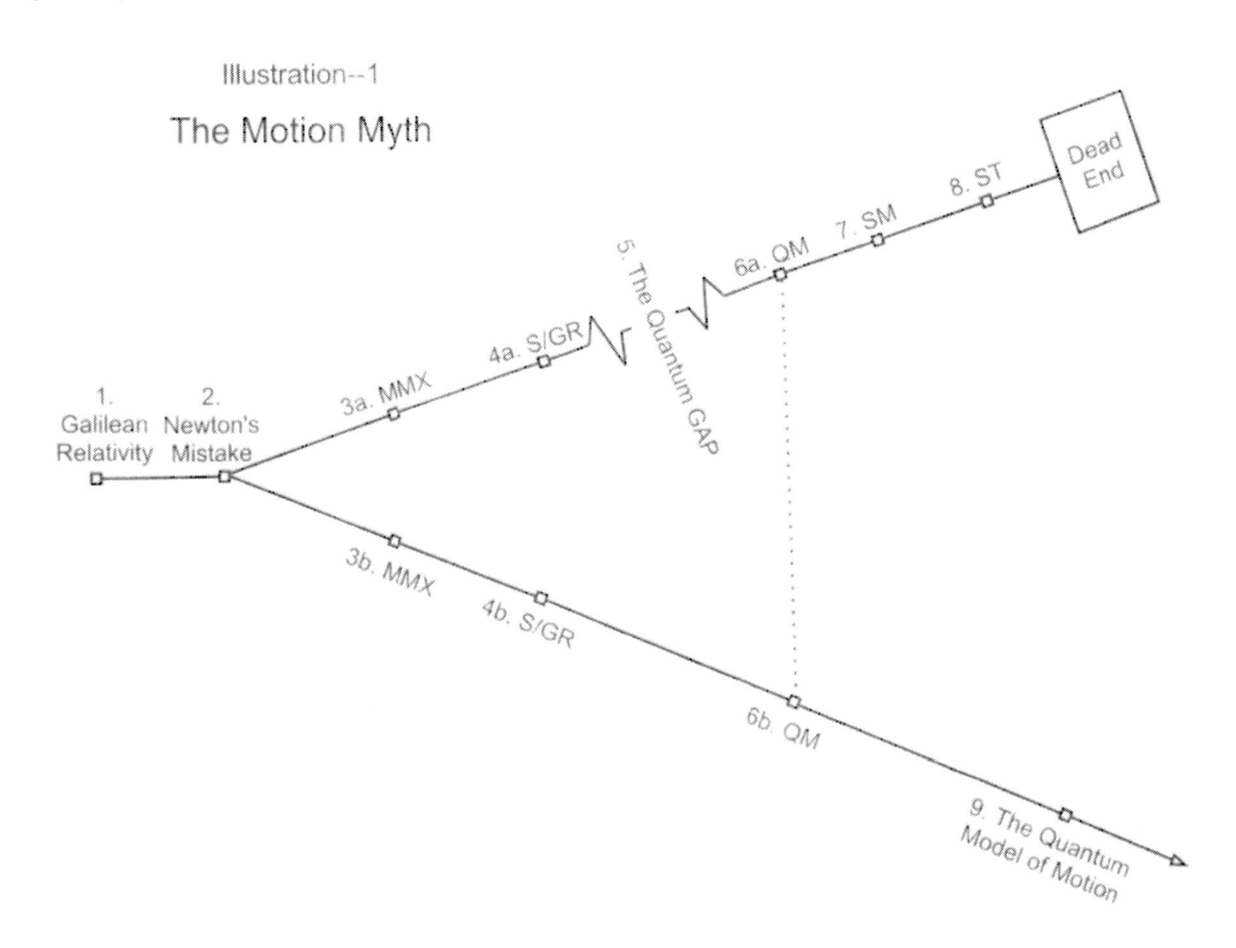

Newton's Mistake was that he failed to recognize that all objects are always in motion, even when they appear to be at rest. This is why he could never explain how gravity really works. He viewed the motion of objects from a whole perspective and not from their individual parts. In doing so, he cemented a precedence that mass just mysteriously moves through space. This is the motion myth. The following illustrates how Newton's mistake perpetuated a paradigm that will eventually lead to a dead end. A new paradigm (The Quantum Model of Motion) is needed in order to explain how gravity really works.

Path (a) represents the path we are currently traveling down. Due to the faulty premise that mass just mysteriously moves through space, it continues towards a dead end that fails to explain how gravity really works. Newton's mistake cemented the fate of this path.

Path (b) represents a new path that must reach clear back to Newton in order to expose Newton's mistake and the motion myth. Then we can reexamine momentum, relativity, and gravity through the lens of a quantum model of motion in order to explain how gravity really works.

1. (Galilean Relativity) From his macro perspective, Galileo concludes that the laws of physics are the same for all inertial frames. Einstein constructs Special Relativity from this premise. Galileo's macro observations fail to question the source and cause of this phenomenon from a quantum perspective. Unfortunately, when quantum physics comes onto the scene, it is not invoked as potentially being able to explain Galilean relativity from a quantum perspective.

2. (Newton's Mistake) From his macro perspective, Newton distinguishes between objects in motion in contrast to objects at rest. From a quantum perspective, there is no difference. All objects are always in motion, even objects that appear to be at rest. Newton's mistake to not recognize this phenomenon perpetuated the motion myth.

3. (MMX) The Michelson-Morley Experiment yielded the contracting nature of mass in motion. It pointed towards a quantum model of motion as the root cause of the contraction of objects in motion through space but became grossly overlooked when Einstein's Special Relativity perpetuated the motion myth with the creation of time dilation and space-time.

4. (S/GR) Einstein's Special and General Relativities are results-based mathematical theories that accurately yield mathematical predictions, but any attempts to link them to quantum physics have failed. Because Einstein was stuck in a macro-motion paradigm, like his predecessors, he failed to see that from a quantum perspective, all objects are always in motion. Ultimately, he failed to explain the momentum of mass, relativity, and gravity from a quantum perspective.

5. The quantum gap was created with the ushering in of quantum mechanics. Quantum mechanics could not be linked to Newton's laws of motion or Einstein's Relativities. Each of these three different viewpoints requires a different mindset to understand. The Quantum Model of Motion is a process-based theory that ties all three of these into a single theory.

6. (QM) Quantum Mechanics deals with the mathematical description of the motion and interaction of subatomic particles but fails to explain why mass as a whole moves through space in the first place. Consequently, it fails to explain the quantum processes involved in the acceleration of mass referred to as gravity. These failures directly stem from Newton's mistake. Quantum mechanics perpetuates the motion myth that mass just mysteriously moves through space.

7. and 8. (SM and ST) The Standard Model and String Theory are built upon a faulty paradigm that doesn't view the source of the motion of mass through space as originating and being sustained from its quantum origins. Once we straighten out Newton's mistake

and the motion myth and explain how gravity really works, then we can see what components of the Standard Model and String Theory fit into and help us better understand the quantum model of motion.

9. The Quantum Model of Motion explains how the motion of all mass through space originates and is sustained from its quantum origins. In essence, it bridges the quantum gap to explain how the laws of motion, relativity, and gravity can all be explained from a single quantum theory.

Newton's mistake remains the main obstacle that prevents us from understanding how gravity really works. The Michelson and Morley experiment provided sufficient insights that could have led to a quantum model of motion that would have exposed Newton's mistake, but Einstein's brilliant misuse of time closed that doorway even to this day.

Part II

The Second Obstacle:

Time

Chapter 4

The Illusion of Time

Sand passing through an hourglass.

Tick…Tick…Tick…Tick…

We've all heard it…when sleep fails us…and we lay awake in the wee hours of the morning listening to its incessant beat. There is never any drum-tick, jazzy cadence…no rhythmic pause here and there…just the never-ending, never-deviating pulse. In the darkness…in the light…it's always there.

Tick…Tick…Tick…Tick…

We place them on our bedside tables…we hang them on our walls…we wear them—these sentinels of existence. They're small and simple in design and function, but magnanimously complex in context and theory. We use them to measure…to plan…even to dictate the day-to-day events that weave the fabric of our lives. Our activities and responsibilities—individually or collectively, locally or globally—hold inferior position to these small, mechanical wonders…clocks!

But, what do clocks do? Sounds like an easy question to answer. We all know what it is…right? It's something we've known since our earliest recollections. We're taught about it from our first weeks of school. We learn to tell it. We're constantly asked about it from passers-by on the street. We watch *it* pass…we squander *it*…we track *it*…we lose *it*…we covet *it*…and we borrow *it*. But, what is *it*…really?

Time.

Aside from the theory of simple forces applied in mechanical movement, is there an actual force called 'time' that moves the hands of the clock? Does it change the numbers on your digital watch…push the earth…or drive the moon? Does it instigate the season's change or cause our cells to degenerate and our memories fade? Does it exist outside the consciousness that defined it…that created it? Is it a physical entity…with shape and form…matter with power to act, and thereby, be acted upon?

No.

Then what is time? In the tangible realm of general physics—the non-quantum world—no equation or computation can stand without it. "Distance over Time" (d/s) permeates all theories of macrophysics. No ball can be thrown, no cannon shot, no train can travel, no rocket launch without including 'time' as a parameter of the solution to any myriad of associated story problems or experiments. It is oftentimes referred to as the fourth dimension, having an equal part in the continuum of space. It is something that simply…*is*.

Or, *is* it?

A study of quantum mechanics, even infantile in depth and breadth, would argue to the contrary. Unlike general or astrophysics, quantum theories abhor the concept of time. In this realm of science…the "sub-atomic" view of the universe…time does not exist. There is no precedence for d/s…no formula or equation containing time as a variable. It is irrelevant. Some may contend as to the express details in reasoning, but, simply put, it is a matter of our inability to gauge with precise clarity the exact location in space of an atom's electron at any given moment in 'time' that compounds the issue. But, does our lack of instruments accurate enough to measure individual electrons and their movements about the nuclei of atoms necessarily corroborate a stand against the existence of time? Our feeling is that our technological accomplishments and creations are irrelevant to the real issue—that of 'time' itself being a purely man-made creation.

Thus, time is a product of human consciousness and serves solely as a humanly intrinsic reference point for the linear measuring of these experiences. It is an arbitrary concept (consisting of limitless range and scope) that attempts to quantify in finite terms…the "infinite"—that very thing which has neither beginning nor end—eternity. 'Time' is indefinable as an absolute entity of the cosmos—holding neither place in space nor exclusive power to change. Take, for example, a brief comparison between the idea of an "inch" and a "second." Both of these units of measurement are simple figments of human imagination. Neither one exists separately from the consciousness that creates and defines them. "Inches" and "seconds" are both arbitrary units of measurement created, defined, and generally accepted by the majority of society to exist and each respectively defines a precise amount of something. These definitions, however, are only inherent to the social/scientific consciousness that obliges them. In other words, they are only valid *here*—on planet earth. There is no universal decree of human preeminence that places our feeble and infantile level of comprehension superior to that of other possible forms of intelligent life…let alone—above that of God.

At this point, one may argue this to be a simple case of semantics… "inches, feet, meters…seconds, hours, days"…what's the difference? They are all units to measure something. Yet, "inches" and "seconds" are finite attempts to capture the infinite. "Inches" attempts to capture a portion of distance, which is a tangible component of space, while "seconds" attempts to capture movement, which is a tangible component of energy. But do "inches" and "seconds" exist outside of conscious thought? Absolutely not! Distance and movement exist outside of conscious thought. But "inches" and "seconds" are human attempts to quantify distance and movement.

So then, what is our definition of time? Time is a measurement of contrasting movements. It is ultimately a measurement of energy, that which cannot be created nor destroyed. In our world, time is the measurement of repeating patterns such as the earth rotating (days) and revolving (years). These patterns become the basis by which all other movements are compared. Then these patterns are divided into smaller but equal parts. We duplicate these natural existing patterns and their divisible portions through mechanical means such as the astutely designed Swiss watch or the Japanese digital wonder.

In the end, there is nothing infinite about the concept of time. It is an arbitrary attempt to measure and cognitively capture contrasting movements. Although one could pose the argument, given society's current paradigms, that a 'second' segments or defines a precise amount of time, the proposal that one could then in turn segment something as broad and infinite in reality as eternity with something as inert and arbitrary as time—having neither form, nor power, nor place in space—is absurd. 'Time' is contrived from human consciousness to define and give order to man's being and the passage of experiences. It serves solely as a reference point—an anchor—for the mortal, linear-thinking mind of man. It does not coexist with the omnipresent state of mass-energy in the universe—mass-energy that can neither be created nor destroyed . . . mass-energy that is eternal in its existence. The concept of time only exists so long as man's consciousness depends on it. In perspective, it is difficult to imagine God…the Alpha and Omega…the Almighty…ruling from His courts on high…immortal…perfect in being and comprehension…possessing absolute and unfathomable light and knowledge…eternal in His existence—being the same yesterday, today, and forever…seated upon His Throne of Glory…wearing a Rolex.

If consciousness ceased to exist, so would the very concept of time with its attendant quantifying units of seconds, minutes, years, and millennia. Mass-energy would still have motion, change, and occupy space—independent of the will of consciousness. We must shed the linear paradigm of time with its accompanying limitations of a "'past, present, and future" if we are to truly understand the eternal, omnipresent state of mass-energy. In other words, 'time' does not exist outside the configuration of the mass-energy of which it purportedly measures.

If mass-energy cannot be created nor destroyed, then in essence, it is timeless. The idea of timelessness has been pondered throughout the ages. Two such philosophers who have pondered the concept of timelessness are Parmenides (515-450 B.C.) and Augustine (354-430).

Parmenides, a philosopher from southern Italy, questioned the reality of a simultaneously existing past, present, and future:

> By employing strict logical argument he produced an interesting idea about Time: all that actually exists is the immediate present. Talk about the past and the future is just talk—neither has any real existence. (Robinson and Groves 1999)

Here is an interpretation of Augustine's struggle to locate his *self* in *time*:

> During the last three years of the fourth century in the Roman city of Hippo in North Africa, a middle-aged man, Augustine, wrote of his life crisis. Among other things, Augustine seems at one point to have lost his 'self.' He could not locate his 'self' anywhere in 'time.' This distressed him. Augustine reasoned thus: The past does not exist—it has left only footprints. Similarly the future does not exist—for it is not yet. But the present also does not exist in a way that one can grab hold of it; it has no 'extension' or 'duration.' By the time one pronounces the word 'present,' the first syllable is forever gone. Hence the crisis: If the past does not exist except in

> memory, if the future does not exist except by anticipation, and if the present does not endure: Where am I? When am I? (Barlow 2007, 1)

Both of these thinkers point out that it is logical to question the nature of time and the actual existence of a past and a future separate from the present. The second thinker even questioned the existence of the present because it had no endurance or duration from one moment to the next. Like the wind, you cannot grasp it without losing its essence. These thinkers were on their way to understanding timelessness as a description of mass-energy from which consciousness flows.

The law of conservation of mass-energy used to be two separate laws up until Einstein. It was the conservation of matter and the conservation of energy. Now it is one law. It basically states that matter and energy cannot be created nor destroyed but can change forms, such as matter becoming energy and energy becoming matter. This essentially says you cannot add to or take away from what already exists.

How did matter and energy come into existence? If matter and energy cannot be created nor destroyed, it would be reasonable to think its existence had no beginning nor will it have an end. The conservation of mass-energy suggests that anything that presently exists has always existed in some form or state, is still existing in some form or state, and will continue to exist forever and ever in some form or state. This idea usually goes beyond the ability of human consciousness to grasp. Our present existence is built on the idea of a beginning and an ending, a birth and a death, such as life, civilizations, planets, solar systems, and even galaxies. Yet, the fundamental substances that make up and allow for the existence of these things are endless in nature, meaning they have no beginning or ending. Mass-energy has always existed and will always exist. It cannot be created nor destroyed. In essence, it has no time.

Then what is time? Time is the incremental measurement of the timeless duration of mass-energy. Time developed as a conscious tool for sequencing movement or changes, giving these segmented parts a pretentious beginning, middle, and end.

The timeless state of mass-energy yields the reality that there is nothing that exists this very moment that hasn't previously existed as some form of mass-energy or will ever cease existing as some form of mass-energy.

Earlier, we mentioned two philosophers who couldn't mentally conceive of an existing past and future separate from the present. Have any recent philosophers or physicists entertained the idea of timelessness?

In the year 2000, an article appeared in Discover magazine. Part of the caption under the title read:

> Imagine a universe with no past or future, where time is an illusion… (Folger 2000, 54)

This sounded like the theory I was developing with a friend prior to the publication of this article. After reading it, I realized that Julian Barbour's ideas of timelessness were completely separate and unrelated to ours. The article did say a few things that really caught my attention.

Julian Barbour was quoted as saying:

> Given what a fascinating thing time is, it's surprising how few physicists have made a serious attempt to study time and say exactly what it is. (Folger 2000, 57)

Also in the article, Don Page, a cosmologist at the university of Alberta in Edmonton is quoted as saying:

> I think Julian's work clears up a lot of misconceptions. Physicists might not need time as much as we might have thought before. He is really questioning the basic nature of time, its nonexistence. (Folger 2000, 61)

I bring up this article that introduces Julian Barbour's book *The End of Time*, to show that the nature of time is still being questioned today.

This brings up the second obstacle to understanding how gravity really works. When the Michelson and Morley experiment left scientists confounded for more than a decade, Einstein's use of time in his relativity theories not only quieted the perplexity of its results, they perpetuated Newton's mistake and the motion myth, leaving quantum and particle physics without the most important key to understanding how gravity really works.

Chapter 5

The Missed Opportunity of the Michelson-Morley Experiment

[From 1887 to 1905, the Michelson and Morley experiment stood between Newton's mistake and the great secret of physics. For eighteen years, the Michelson and Morley experiment perplexed the scientific community. No one could correctly explain the cause for the contraction of mass in the direction of motion. With quantum mechanics in the horizon, this dilemma could have spurred the connection between the motion of mass through space and quantum physics. In 1905, however, Einstein's brilliant calculations mortally wounded the potential unveiling of the great secret of physics, keeping it obscure until now.]

In the late 1800's, scientists believed *ether* was a motionless substance through which the earth moved.

Illustration--2

Earth/Ether Model

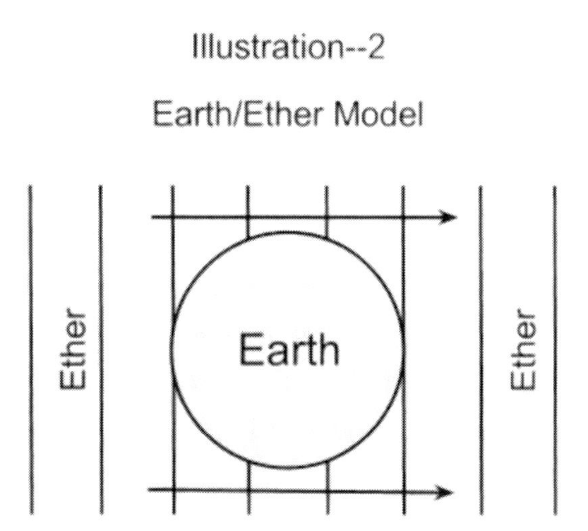

The Michelson and Morley experiment was specifically designed to measure the movement of thc carth through this substance that many scientists believed permeated space. Just as energy travels through water in waves, scientists believed there needed to be a substance permeating space that light waves propagated through. Michelson and Morley hypothesized that the movement of the earth through this ether could easily be detected by measuring the interaction of light directed into the Michelson and Morley apparatus. As light entered into the apparatus, it would be separated into two separate beams by using a mirror and a clear window right next to each other. Then the divided light beams would travel at right angles to each other, bounce off mirrors, and be brought back together.

Illustration--3

The Michelson and Morley Apparatus

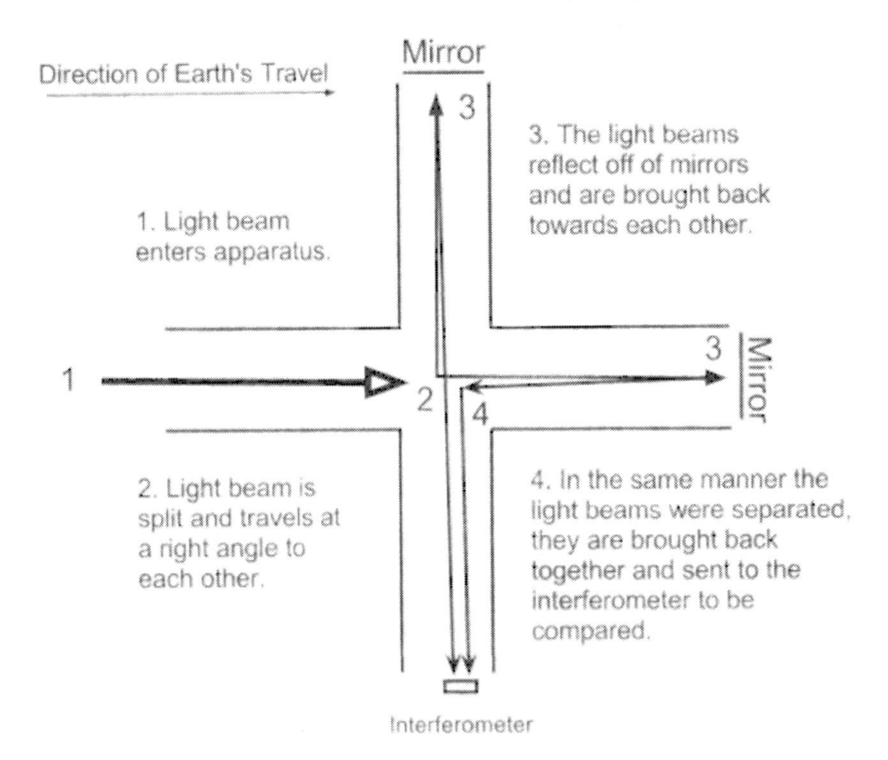

After the light entered into the apparatus and the mirrors were adjusted, they turned the apparatus slowly while measuring the results through an interferometer, an instrument that measures wave interference. After the split beams were brought back together, according to calculations, the light should have traveled different distances, resulting in interference patterns. This means that the wavelengths of the light that were divided and brought back together should no longer be lined up with each other but slightly shifted due to traveling different distances. The degree of the interference would indicate the speed of the moving earth through the stationary ether. The key to this experiment was that the apparatus was resting on the orbiting earth, which travels around 19 miles per second through space. As the earth, with the apparatus attached to it, orbited the sun, the apparatus would slowly be turned. The rotation of the apparatus changed the angles the light moved through the ether.

The results of this carefully crafted experiment surprised everyone. The expected interference patterns were not detected. At this point, Michelson and Morley believed their experiment failed and went back to the drawing board. When they redid their experiment and obtained the same outcome, scientists attempted to explain the validated results. Both George Fitzgerald and Hendrik Lorentz independently came up with a mathematical contraction formula to explain the contraction of the apparatus needed to understand the null results (no interference patterns) of the experiment.

George Francis Fitzgerald thought the apparatus could have contracted as it moved through the ether by external pressure from the ether on the apparatus. This, he thought, would explain the null results of the experiment. He developed a contraction equation that accounted for the measured null results.

> Trying to account for Michelson's failure to find any movement of the earth in relation to the ether, Irish physicist George Francis Fitzgerald suggested that the measuring instruments Michelson used had contracted slightly and distorted the reading. He then produced equations showing that matter contracts in the direction of its motion, the contraction increasing as the speed increases. The Fitzgerald contraction, as this phenomenon is known is 'exceedingly small in all ordinary circumstances… If the speed is 19 miles a second – the speed of the earth around the sun—the contraction of length is 1 part in 200,000,000 or 2 ½ inches in the diameter of the earth.' (Brian 1996, 66)

Hendrik Lorentz also hypothesized that the apparatus somehow contracted in the direction of movement to account for the null results of the experiment. He considered electrical and magnetic fields within the apparatus to account for the contraction. He, too, devised a contraction equation, which is referred to as the Lorentz transformation. The Lorentz transformation formula was used by Einstein for his Special Theory of Relativity.

> Dutch physicist H. A. Lorentz stated that a flying charged particle foreshortened in its direction of travel would increase in mass. Einstein, in turn, applied Lorentz's equations, known as the Lorentz transformation, to all objects, including clocks and measuring instruments. Einstein showed that objects moving great speeds and over vast distances decreased in size and increased in mass. (Brian 1996, 66)

Lost in Einstein's1905 Special Relativity mathematical equations is a quantum explanation of how the Michelson and Morley apparatus contracted. Both Fitzgerald and Lorentz independently concluded that the apparatus physically contracted and that the degree of the contraction would change at different speeds. In other words, could every atom of the apparatus proportionately foreshorten in the direction of travel while increasing in mass? Only quantum physics could potentially explain the contracting apparatus from an inside-out perspective. During this time period, quantum physics was still in its prenatal stage. Partially for this reason, Einstein's new Special Theory of Relativity completely bypassed the role of quantum physics in the contraction of moving mass. Even later in Einstein's life, years after the development of quantum mechanics, he never explains the quantum process of why object moving at great speeds decreased in size and increased in mass. Like Newton, who discovers gravity but retards its eventual explanation, Einstein, standing on Newton's shoulders, uses mathematics to more accurately predict the results of gravity while perpetuating Newton's mistake and the motion myth, further hindering a quantum explanation of gravity. If you don't understand that all mass is always in motion through space, then you can't surmise what might be happening on a quantum level with increases and decreases to its speed.

Chapter 6

Einstein's Misuse of Time

[The biggest proof of Einstein's real blunder is the notepad and pencil he used as he was approaching death, scratching down equations in a futile attempt to configure a unified theory. In the end, Einstein's relativity theories didn't bring him closer to a unified theory. They actually erected impenetrable barriers to that desired end. Time dilation and space-time are products of the motion myth paradigm Einstein was stuck in when he developed his Relativity theories.]

The purpose of the Michelson and Morley experiment was to measure the ether wind with the split light beams. They didn't set out to prove or disprove whether or not the ether existed. That is why the results of their experiment greatly surprised them. With no permeating ether as a base to compare the constant speed of light, Einstein theorizes that the speed of light is the same for all moving objects. He based this assumption on Galileo's observation that the laws of physics are the same for all inertial frames. According to Einstein, James Clerk Maxwell's constant speed for all electromagnetic waves must be constant for all inertial frames. By carefully crafting time into the Lorentz transformation, Einstein showed how this was mathematically possible. Einstein once said the following in referring to a mistake made by Max Planck:

> The main thing is the content, not the mathematics. With mathematics, one can prove anything. (Brian, 1996, 78)

This is where standing on Newton's shoulders tainted his reasoning when it came to the motion of objects through space. He still wanted to treat the motion of objects as a whole instead of from their individual parts working together. He outright skipped the question of why objects have motion in the first place and built off the faulty paradigm that mass just mysteriously moves through space. From this paradigm, he postulated that the speed of light is the same for all moving objects.

Using the Lorentz transformation and the concept of time, he built his elaborate theory to demonstrate how the speed of light is the same for all moving objects. According to Einstein's Special Relativity Theory, the faster an object moves through space, the slower its time frame. This mathematically allows the speed of light to be the same for all moving objects, such as a walking person, a racing car, a speeding rocket, or an orbiting planet. This space-time physics masked the need for a quantum model of motion.

Einstein made the same fundamental error when explaining gravity with his General Theory of Relativity. His space-time solution failed to address why mass moves through space in the first place. His theory provided mathematical accuracies without addressing quantum causes. His external emphasis on objects as a whole perpetuated Newton's mistake instead of exposing it. For this, he paid the heavy price of not being able to explain gravity from a quantum perspective.

In the end, both relativity theories fail to answer a basic fundamental question of physics: *How does mass move through space in the first place?* Had Einstein asked this question, he might have derived that all objects are always in motion and are never at rest. For what appears to be "at rest" is really the perpetual motion of mass under the spell of terminal velocity and gravity. Working together simultaneously, they create the illusion that objects can obtain a state of rest within an inertial frame. Insight into this reality would have pointed towards the underlying role of quantum physics in the motion of mass through space. Instead of exposing the motion myth, Einstein brilliantly perpetuated it. Einstein's Relativities are mathematical masterpieces that make accurate predictions but fail to incorporate quantum physics into its proper role of explaining motion, relativity, and gravity.

For Einstein, as with Newton, objects just mysteriously move through space. There is no quantum explanation for the cause and changes of inertia. In a quantum model of motion, all motion and changes in motion are explained in detail from a quantum perspective. The biggest difference between Einstein's Relativities and a quantum model of motion is that a quantum model of motion explains how gravity really works from a quantum perspective, Einstein's unfinished project.

In summary, Einstein's use of time in his Relativity theories provided a mathematical substitute for quantum processes that could not be perceived within a motion myth paradigm. A quantum model of motion exposes Einstein's misuse of time and ends the need for time dilation and space-time to explain relativity and gravity.

Einstein's real blunder was not exposing Newton's mistake and the motion myth.

Chapter 7

Quantum Physics Misguided

"I like to think that the moon is there even if I am not looking at it"
Albert Einstein

The profound influence of Newton's mistake may have had its greatest detrimental consequence at the inception of quantum physics. At the time when scientists were exploring the composition of an atom and its role as the building block of all matter, they should have also considered the role of quantum physics in the motion of mass through space. With the scientific belief that all mass originates on a quantum level, is it not logical to think that the motion of mass through space also originates on the quantum level?

The unsuccessful equations Einstein jotted down and scratched out with his dying hands testify that the archaic belief that mass just mysteriously moves through space is grossly insufficient in a science enlightened by quantum physics. This faulty paradigm that mass just mysteriously moves through space is still the predominant paradigm that is unknowingly weaved into all present theories, including the Standard Model. The Standard Model attempts to explain the origins of mass but grossly fails to explain how gravity really works. You can't have the one without the other. You can't explain the origins of mass without explaining how gravity really works. Is not gravity the unifying mechanism upon which all of physics and life evolved?

Two of our greatest physicists, Newton and Einstein, viewed the motion of objects through space from a whole perspective and not from their individual parts, perpetuating the motion myth. Quantum physicists, standing on their shoulders, also failed to see the underlying role of quantum physics in the motion of mass through space. This erroneous paradigm that mass just mysteriously moves through space has now been passed down to us and continues to stifle our ability to see the underlying role of quantum physics in the motion of mass through space and the underlying role of quantum physics in relativity and gravity.

In summary, quantum physics was misguided from its inception when quantum physicists naively ignored the underlying role of quantum physics in the motion of mass through space. As a result, gravity remains a mystery. If we want to understand gravity from a quantum perspective and how it really works, we must come to the conclusion that all objects are always in motion through space and that the individual atoms making up those objects are the source and cause of that motion. Then we will begin to clearly see from an inside-out perspective how quantum mechanisms majestically orchestrate the perpetual motion of all mass through space. This is the grand key for understanding quantum gravity.

At some point, we will all wonder why it took so long for the scientific community to surmise the absolute need for a quantum model of motion in order to explain how gravity really works.

Part III

The Third Obstacle:

Finding the Big Picture: The Quantum Model of Motion and The Energy Cycle

[The Quantum Model of Motion is a philosophical blueprint to aid science in explaining how gravity really works. The quantum model of motion completes the border of the gravity puzzle, the unifying mechanism upon which all of physics and life have evolved. Others can come along to straighten out the middle pieces.]

Chapter 8

The Quantum Model of Motion Overview

The Quantum Model of Motion answers the question that should have been asked centuries ago: *How does mass move through space in the first place?* As already emphasized in this book, great physicists such as Galileo, Newton, Einstein and beyond have taken for granted that objects just seem to move through space with now accountable science explaining how or why this happens. These reverenced names proceeded to explain relative relationships between moving bodies with laws and theories, but notwithstanding their enlightened contributions, the simple yet profound movement of mass through space whispers: *What is the science of motion?* This question delicately holds the missing link between Newtonian physics and quantum mechanics and is the keystone to a unified theory that explains how gravity really works.

Even in the declining years of his life as he struggled to compose a unified theory, Einstein still missed this elusive yet simple truth that evaded his predecessors. Simply put, mass does not mysteriously move through space. There is a scientific explanation that accounts for the perpetual momentum of all masses and changes to that momentum. The quantum model of motion unveils the science of motion from an atomic perspective. Just as all objects of any size are made up of atoms, the momentum of all objects, including large massive objects such as planets, solar systems, and galaxies, are initiated and sustained at the atomic level.

It is important to establish *the atomic model of motion* before delving into *the quantum model of motion*. The atomic model of motion deals with the science of motion for an individual atom, taking into consideration differing atomic numbers. The foundation for the atomic model of motion is that the protons and neutrons of the nucleus of an atom are responsible for that atom's perpetual momentum through space. Any change in an atom's momentum yields an accompanying change in the energy level of each proton and neutron of that atom.

The quantum model of motion deals with the science of motion for bonded atoms in the form of mass. Since mass is made up of atoms, the momentum of mass can easily be described as the synchronized movements of individual atoms. This is the starting point for the quantum model of motion. The quantum model of motion consists of the following four four comprehensible parts:

1. Quantum momentum is the equilibrium state of momentum where energy is neither absorbed nor emitted. This is the reason that an object in motion stays in motion.

2. Quantum adjustments are changes in momentum (acceleration, deceleration, and changes in direction) that are accompanied by the absorption or emission of energy to exactly compensate for those changes. The science of motion is as exact as the science of chemistry. All changes in momentum can be explained by the addition or subtraction of energy.

3. Quantum relativity explains Galilean relativity from a quantum perspective. The sustaining energy driving all inertial frames is measurable and accountable on a quantum level using an understanding of quantum momentum and quantum adjustments.

4. Quantum gravity, the acceleration of mass that we call gravity, is nothing more then the absorption of energy that accelerates already moving mass in the direction of absorption. The result of absorbed energy is acceleration. The effects of increased energy within mass due to the effects of gravity are easily observable.

The quantum model of motion amalgamates the physics of Galileo, Newton, Maxwell, Einstein, and Heisenberg into a single, complete, and comprehensible theory that bridges the gap between macro- and microphysics. It is the keystone for understanding how gravity really works.

Chapter 9

The Atomic Model of Motion

In order to explain the nuts and bolts of gravity, a modified model of an atom is needed. Both the Bohr model and the quantum mechanical model fail to take into consideration the role of the atom in the momentum of mass. Any model of an atom is incomplete without explaining its perpetual momentum as an integral part of its very structure. The energy making up an atom is responsible for that atom's momentum through space. In other words, atoms don't just mysteriously move through space. The components or parts that make up an atom simultaneously account for its momentum. At all times, atoms maintain momentum even as they combine in the process of fusion to make larger atoms, or bond in the form of molecules and compounds, or even when they reside in a mixture. All atoms maintain momentum even if they appear to be at rest, such as on the earth's surface. As the strong force holds an atom together, the atom's ability to absorb and emit energy regulates its speed and direction through space. As speed or direction change, so does the gradation of energy making up the atom.

The basis for the atomic model of motion is actually quite simple to visualize. Think of each individual proton and neutron as packets of energy moving through space. Unlike photons, which move through space at the same fixed speed, a nucleus consisting of protons and neutrons moves through space at varying speeds in direct proportion to the amount of energy making up each proton and neutron within the nucleus. The following will help explain how this might be possible. Think of a particle moving through space in a circular-spiraling momentum. As the energy making up the particle moves through space at a fixed speed, the size of the particle's circular-spiraling momentum determines the linear speed at which the particle as a whole moves through space. A larger circular-spiraling momentum moves at a slower linear speed through space than a smaller circular-spiraling momentum. You can change the linear speed through space by changing the size of the circular-spiraling momentum. For example, the same length of strings can have different linear lengths by placing them side-by-side in wave-like patterns. By changing the wavelength and amplitude of one string, you change its overall linear length in comparison to the linear length of the other strings.

In the atomic model of motion, each proton and neutron of the nucleus of an atom is the driving force for the motion of that atom through space. The strong force binds the independent momentum of each proton and neutron to simultaneously create the structure and momentum of that atom. It is like a team of horses yoked together. Their individual movements are synchronized. The force holding together the momentums of the protons and neutrons is stronger than the independent momentums of each proton and neutron. This is like the yoke that keeps the individual movement of each horse bound together. At some point when the individual momentums of each proton and neutron becomes stronger than the force holding them together, the nucleus breaks apart into smaller atoms as in the case of nuclear fission.

Atoms bond together forming various types of molecules of differing combined masses. The atoms of these masses move together like synchronized swimmers creating the momentum of these masses through space. The bonds that hold atoms together are weaker than the strong force that hold the protons and neutrons of a nucleus together and are more susceptible to breaking apart but are also more easily joined together. They break apart when the individual momentums of each atom become stronger then the bonds holding them together such as when a mirror is dropped on a hard tile floor.

Electrons that orbit the nucleus of an atom continually alter their movements through space in order to continue to orbit the nucleus as the nucleus moves through space. They fully participate in the synchronized dance with the protons and neutrons that they orbit. I wouldn't be surprised if there was an energy dance that transpires between the nucleus of the atom and the electrons orbiting it when the need arises, such as when changes in momentum occur. As electrons continuously absorb and emit photons in order to change shells, it is possible that when the momentum of protons and neutrons are altered, energy is transferred between the electrons and the nucleus of an atom to exactly compensate for the energy changes to the protons and neutrons. As the momentum of protons and neutrons increases, energy could be absorbed into the nucleus of the atom from the electrons. As the momentum of protons and neutrons decreases, energy could be emitted from the nucleus of the atom to the electrons. The transfer of energy between electrons and the nucleus of an atom to exactly compensate for changes in speed and direction of the protons and neutrons that make up the nucleus of an atom coincides with what we already know about atoms. The main point is not how energy is transferred into and out of the nucleus of an atom but that energy is transferred into and out of the nucleus of an atom to exactly compensate for momentum changes.

The atomic model of motion is the perpetual momentum of an atom through space. Its varying speeds through space correlate exactly with energy that is absorbed into or emitted out of its nucleus. I cannot explain how the quarks with their spins, along with other subatomic particles, work together to maintain the momentum of an atom through space. Or the exact process of how energy is absorbed and emitted by every proton and neutron of an atom in order to exactly compensate for momentum changes. It is like watching the hands move on a mechanical watch without knowing specifically how the gears interact to create the methodical movements that yield the time. I can observe the movements of the hourly and minute arms without knowing the specifics about the gears, but I do know that the movements of the hands of a clock would not be possible without the movement of gears. The gears represent inner pieces of the quantum puzzle where others have greater specialty. The purpose of this book is to connect the border pieces together so that the inner pieces have somewhere to go and can complete the picture.

In summary, the atomic model of motion asserts that every proton and neutron of every atom is composed of confined energies that simultaneously account for the structure as well as the perpetual momentum of every atom through space. The atomic model of motion sets the stage for the quantum model of motion. Every minute detail of the quantum model of motion is built on the foundation of the atomic model of motion. The atomic model of motion is the Rosetta Stone for explaining the mystery of momentum, relativity, and gravity just as plate tectonics is the Rosetta Stone for explaining geological changes of the earth's surface.

Chapter 10

Quantum Momentum

The Underlying Principle Sustaining Uniform Motion

This chapter gives a philosophical overview of quantum momentum. Chapter 11 on quantum adjustments will explain the science of motion sustaining quantum momentum.

Isaac Newton's first law of motion states that an object at rest stays at rest and an object in motion stays in motion with the same speed and in the same direction unless acted upon by an unbalanced force. What is the difference between an object at rest and an object in motion? Nothing. Had Newton recognized the motion myth paradigm, he would have clarified that there is no difference between a body at rest and a body in motion. Both objects are in motion. It is just that the body at rest is sharing the same motion as the frame it appears to be resting on. All objects are always in motion. When acted upon by uneven forces, they can either accelerate or decelerate, but at no time do they ever stop being in motion. The perpetual motion of all mass can only truly be explained or understood from a quantum perspective. For example, why do asteroids in the asteroid belt continually orbit the sun without ever running out of fuel? You just push the rock and because of the magic of inertia, it continues to move through space at the same speed and in the same direction until a force acts upon it again? In the day and age of quantum physics, is the idea that mass just mysteriously moves through space acceptable science?

If objects don't just mysteriously move through space, then what is the source and cause of their motion? The answer is quite simple: The atom. All atoms are always in motion. The equilibrium state of any atom is uniform motion. Any change to its uniform motion is accompanied by the absorption or emission of energy since a change in motion is a change in the energy driving that motion. The atom is the never-ending energy that keeps all mass in perpetual motion. This is how planets continuously orbit the sun without having to be rewound or refueled. Even objects appearing to take a break on earth's surface still cruise at nineteen miles per second as gravity yields them earthbound. Turn off gravity and these objects are no longer attached to the earth's surface. Give them a little nudge and they will float off in uniform bliss until an interaction with some form of energy redirects their route or changes their speed.

This is why Newton's mistake adversely affected the development of physics. The very assumption that an object is at rest, that it is not exhibiting motion independent of the source upon which it appears to be resting, negated the potential of pinpointing quantum energy as the source of its motion. In the emergence of an atom, energies form the atom's structure while simultaneously engaging its momentum through space. The two, (structure and momentum), are inseparable. The energy making up the atom perpetually drives its sustained uniform motion until unbalanced forces act upon it. Since atoms make up mass, the inertial momentum of mass is simply the synchronized uniform motion of each individual atom making up that mass.

The mystery of why an object in motion stays in motion can now be unveiled as the synchronized momentum of bonded atoms making up that object. The momentum of synchronized atoms in the form of a mass continues in uniform motion until a force acts upon them. The force alters the energy and structure of each atom. This initiates changes in the speed and or direction of the mass as a whole.

At first, the idea that the momentum of mass through space is the synchronized momentum of the collection of atoms making it up may be difficult to visualize, but as the pieces to the quantum model of motion come together, the process of visualizing mass moving through space from this insight becomes very easy and natural.

What does make this concept difficult to grasp is we continually see mass at rest on the earth's surface with no apparent motion whatsoever. The computer I am typing on does not appear to have any momentum. This appearance of mass at rest is the same illusion that fooled Galileo and Newton and is still fooling physicists to this present day. The computer is in a continuous state of acceleration towards the earth and only appears motionless because it is temporarily experiencing terminal velocity. Its uniform motion is riding on the uniform motion of the earth due to the effects of gravity. If the effects of gravity could temporarily be shut off, the momentum of the computer through space would be more apparent. It would continue to move through space at the same momentum of the earth without being connected to the earth because it would no longer be accelerating into it. If you gave it a push, it would slowly drift away from the earth, accentuating its own momentum through space, the synchronized momentum of the atoms from which it is made.

A good way to visualize quantum momentum is to think about objects in momentum in the space shuttle while it is orbiting the earth. When the space shuttle is in a free-fall, the independent momentum of all objects can clearly be observed. When an astronaut pushes an object, such as a bag of water, the bag of water continues in the same direction until it bumps into a wall of the space shuttle. As it runs into the wall, its momentum shifts to a new speed and direction only to continue in that path until a force acts upon it again. It would continue its momentum uninterrupted even if the space shuttle were to suddenly vanish from around it. Objects that the astronauts place in the air next to them—such as a toothbrush—stay in the same place. This is because they have the same momentum through space as the astronaut. Their momentums are synchronized, moving at the same speed and in the same direction.

The atoms that makes up any mass share an order of movement that orchestrates that mass's movement through space. Just as the momentum of an atom is never at rest, mass is never at rest, either. The perpetual motion of mass, as a collection of synchronized atoms, is either experiencing uniform motion, terminal velocity, (which is still motion, it just wants to accelerate its motion but can't), acceleration, or deceleration. At no point though do the atoms in motion cease to be in motion. Mass that appears to be at rest is just sharing the same motion with the object it appears to be resting on.

An example of the shared uniform motion of separate masses not connected to each other is objects within a car but not connected to it. As the vehicle accelerates, all the objects within the car go through internal changes to accommodate for the acceleration of the vehicle they appear to be resting in. At each new speed, all the objects in the car obtain the same uniform or quantum momentum as the car. When the car crashes into something, as its momentum is instantly changed, each object continues in their own independent momentum until they crash into something such as the back of the seat, the dashboard, or the windshield. Each object has its own unique inertial momentum independent of all the other objects, including that of the car. That is why we wear seat belts to physically attach our momentum with the momentum of the vehicle in which we are riding.

Quantum momentum is the inertial momentum of mass through space, which is caused by the synchronized momentum of the atoms from which it is made. The speed and direction of mass through space remains unchanged just as Newton observed until an unbalanced force or energy acts upon it. Then the speed and direction of that motion are altered, but it never stops or rests. Newton's mistake was his failure to realize that all mass has motion, even objects that appear to be at rest. The inertial momentum of all mass, even objects that appear to be at rest, is an important key to understanding how gravity really works.

Chapter 11

Quantum Adjustments: Changes in Momentum

Quantum momentum is the inertial momentum of mass;
Quantum adjustments take place with changes in momentum.

The momentum of mass does not change—a brilliant call by Newton— unless acted upon by a force. The result of that force is a change in momentum. A change in momentum is proportionately accompanied by the absorption or emission of energy. Exactly how the protons, neutrons, and electrons work together with other energies to regulate changes to momentum is not fully known. The following represents possibilities to explain the chain reaction of momentum changes. As changes in momentum occur, be assured that energy transfers also transpire. The science of movement is exact; the knowledge of exactly what happens and how it happens is in the developmental stage.

The Absorption and Emission of Energy

The absorption and emission of photons to account for the shifting of electrons between the shells of an atom is common knowledge within the circles of acceptable science. This demonstrates an atom's plasticity to accommodate energy input and output. In describing changes in momentum, we take this plasticity a step further by pointing out that an atom's nucleus is also in a constant flux of change. It absorbs and emits energy to accommodate changes in momentum.

If an atom's momentum increases, the atom's nucleus will absorb energy that exactly correlates with that atom's new momentum, and if an atom's momentum decreases, the atom's nucleus will emit energy that exactly correlates with that atom's new momentum. The absorption and emission of energy has the purpose of reestablishing equilibrium within the atom as it goes through momentum changes. When the atom is in a state of equilibrium, that atom is in uniform motion and will maintain that motion until acted upon by a force or energy.

During momentum changes, each proton and neutron of that atom independently absorbs or emits the exact amount of energy to compensate for the momentum shift. This regulates the momentum of an atom through space, whether the atom is by itself or is part of a collection of atoms in the form of a mass. I believe electrons fully participate in this dance to aid in the absorption and emission of energy in and out of the nucleus. The main point is that the atom's structure is of such that energy is absorbed or emitted to exactly compensate for changes in momentum. I refer to this as quantum adjustments—energy adjustments into and out of the nucleus of an atom that exactly correlates to increases and decreases in momentum.

Momentum Patterns and Energy Levels

The momentum pattern and energy level of an atom are two sides of the same coin that help explain changes to the continuous momentum of atoms through space. During the formation of an atom, protons, neutrons, and electrons unite in a synchronized dance to form and maintain the structure and motion of that atom through space. This structure, the atom's *momentum pattern*, describes the atom's perpetual momentum through space. The *momentum pattern* of an atom remains constant unless unbalanced forces act upon it. For this reason, I refer to an atom's uniform motion through space as a *momentum pattern* because this pattern continues until it is disrupted. As an atom's momentum changes, energy is proportionately absorbed or emitted to exactly compensate for these changes. Quantum adjustments in the form of energy absorption and emission are necessary to exactly compensate for disruptions to an atom's *momentum pattern*.

The best way to describe a *momentum pattern* is to go back to the horse example. Think of one set of a proton and a neutron as a pair of horses yoked together side by side walking at the same speed and in the same direction. Their combined movement is their momentum pattern. If we added additional pairs of horses to the first pair, so that each pair was lined up like a team of dogs pulling a sled, the combined movement of each pair of horses would be the momentum pattern of the combined team. Notice that no matter how many sets of horses we yoke together, each individual set of horses move at the same speed and in the same direction as all the other sets of horses, but with each additional set of horses added, the overall pulling power increases. Thus, the momentum pattern describes the speed and direction of the yoked horses, notwithstanding the number of sets.

Like the yoked horses pulling the sled, the nucleus of an atom could be described as combined energies that simultaneously make up an atom's structure and momentum through space. It is this energy, the energy that makes up the protons and neutrons of the nucleus of an atom, which accounts for the motion of that atom through space. Like electrons, protons and neutrons can absorb and emit energy. This regulates their momentum through space. As the linear speed of an atom increases, energy is simultaneously absorbed into every proton and neutron of that atom. As the linear speed of an atom decreases, energy is simultaneously emitted from every proton and neutron of that atom.

The speed of any atom through space is determined by the *energy level* of each proton and neutron that makes up its structure, its *momentum pattern*. As a force increases the speed of an atom, energy is absorbed into each proton and neutron. This increases the *energy level* of each proton and neutron of that atom. The same atom will now have an overall higher *energy level* per proton and neutron. Conversely, as a force decreases the speed of an atom, energy is emitted from each proton and neutron. This decreases the energy level of each proton and neutron. The same atom will now have an overall lower *energy level* per proton and neutron. The *energy level* of an atom is in direct proportion to the energy level of each proton and neutron of an atom.

Energy level could be used in two different ways. There is the overall energy level of an atom and the *energy level* of each pair of a proton and neutron of an atom. There is an important distinction. Compare an atom with one proton and neutron pair to an atom with five pairs of protons and neutrons. If these two different atomic size atoms have the same momentum through space, then the *energy level* of each pair of protons and neutrons is equal, but the overall energy level or amount of energy will be greater for the atom with five pairs than the atom with one pair. It will have five times the overall energy, even though the atoms move through space at the same speed. The atom with the five pairs will have more inertia or resistance to change than the atom with one pair. Throughout the rest of this book, when I refer to the *energy level* of an atom, I am referring to the energy level of one pair of a proton and neutron. This is the *energy level* that describes its linear speed through space, notwithstanding the atom's atomic number.

The protons and neutrons ability to absorb and emit energy allows for an atom to adjust its momentum when a force acts upon it. This holds true for each atom, notwithstanding the number or protons and neutrons in its nucleus. As the atom increases in speed or changes direction, energy is absorbed equally into every proton and neutron of the nucleus of that atom. As the atom decreases in speed, energy is emitted equally from every proton and neutron of the nucleus of that atom. This is how the energy level and momentum pattern are inseparably connected. A change in the momentum pattern of an atom initiates a simultaneous change in the atom's energy level, the energy level of each proton and neutron. By the same token, a change in the energy level of every proton and neutron of an atom constitutes a corresponding change in its momentum pattern, changing the speed and or direction at which that atom moves through space. This last part is an important key in explaining quantum gravity.

The horse example shows how the momentum pattern and energy level are actually a single process. The *energy level* is the speed at which each horse moves. When yoked together, each horse maintains the same energy level. This is also the momentum pattern, not only for each horse, but also for all the horses combined. The speed of each individual horse is also the same speed of the entire team. Imagine you have a team of six horses, with two horses yoked side by side, with another set yoked behind them, and the last set yoked behind them. The speed that each horse moves is the momentum pattern for that whole team. They all move at the same speed. When you apply a force, such as a whip to speed them up, or a pull on the reins to slow them down, you simultaneously change the energy level (speed) of each horse and the momentum pattern (speed) of the whole team. The energy level and momentum pattern for one horse is the same for all the other horses. The energy level and momentum pattern for one set of a proton and a neutron is the same energy level and momentum pattern for all the other sets of a proton and a neutron within the same atom. When you change the momentum pattern, the energy level simultaneously changes. And vise versa, when you change the energy level, the momentum pattern simultaneously changes. They are two sides to the same coin.

Mass is a collection of synchronized atoms moving through space as the momentum pattern of each atom moves in concert with all the other atoms of that mass. Mass maintains a constant speed and direction until there is a disruption to the momentum pattern of each atom that makes up mass. As the momentum patterns of the atoms of a mass are disrupted, energy is absorbed or emitted from the mass, from each individual proton and neutron of each atom. This reestablishes equilibrium within each atom within the mass to accommodate its new speed and direction through space.

When two masses collide, the *momentum patterns* of the protons and neutrons of the atoms of each mass are disrupted. As the *momentum patterns* are altered, the accompanying *energy levels* of the protons and neutrons of the atoms of each mass simultaneously adjust to accommodate the new momentum patterns. An absorption or emission of energy to exactly coincide with the degree that the *momentum patterns* were altered transpires. As equilibrium is restored, the masses continue in their new momentums until a force or energy acts upon them again.

In summary, the momentum pattern and energy level of each proton and neutron that makes up the nucleus of an atom regulates the speed and direction of that atom through space. When momentum changes occur, quantum adjustments transpire to reestablishes the equilibrium of quantum momentum. Since mass is the synchronized movement of individual bonded atoms, a momentum change in mass is a momentum change for every atom that makes up mass. Every atom experiences a quantum adjustment. As there is *no nothing for something* in the cause and effect science of physics, changes in the momentum of mass must be accounted for by corresponding energy changes.

Chapter 12

Quantum Relativity

Quantum relativity is an explanation of Galilean relativity from a quantum perspective.

How can the same falling object literally travel multiple different distances?

Galilean Relativity

Galileo learned that if a person were in a closed cabin of a ship moving at a uniform speed, he wouldn't be able to tell his relative motion to land by an appeal to the laws of physics. If he dropped an object, it would fall straight down. If he threw it up in the air, it would go straight up and straight down. In all, there was nothing he could do to determine his relative speed to land by applying what he knew about the laws of physics.

On the surface, the laws of physics appear to be the same for all inertial fames, but a deeper look into the microphysics of relativity reveals differences on a quantum level for the same mass in differing inertial frames. The reason the laws of physics are the same for all inertial frames can only be understood from a quantum perspective.

If person A, traveling on a train holds a beanbag four feet above the floor and drops it to the floor of the cabin wherein he is riding, he will measure the beanbag to have fallen four feet. If person A decides to drop the same beanbag out of the moving train's window four feet above the ground, person B, standing on the stationary earth relative to the moving train, will measure the distance of the falling beanbag to fall more than the four feet as observed by person A, who sees the beanbag fall straight to the ground. For person B, the beanbag will not only fall four feet to the ground as measured by person A, it will also fall at an angle proportionate to the speed of the train. For this reason, each observer will literally measure the beanbag falling different differences. (See illustration—4). This is classic Galilean relativity, but it brings up an interesting dilemma: *How can the same falling object literally travel multiple different distances?* The answer to this question will be found in the *momentum pattern* and *energy level* of each atom involved. (Note: Since momentum pattern and energy level are two descriptions of the same process, I use them interchangeably to avoid the repetition of saying them both every time. I more often use momentum pattern to focus on the speed of an atom through space, and energy level to compare atoms in different inertial frames.)

Illustration--4

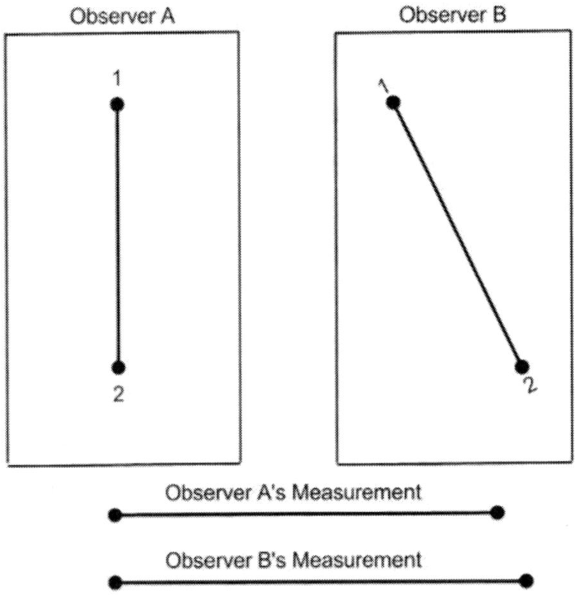

Observer A

Observer B

1

2

1

2

Observer A's Measurement

Observer B's Measurement

With each incremental increase of speed to the beanbag as it travels in a moving train—relative to earth's inertial frame—the more overall energy the beanbag acquires. For example, if person A drops a beanbag four feet into a pile of sand outside of a non-moving train, the beanbag will displace a certain amount of sand upon impact. The energy acquired by the falling beanbag due to the effects of gravity is transferred to the sand at impact. Now, if the train is moving and person A drops the beanbag the same four feet so it hits the same sandbox that is outside of the train, falling at an angle due to the movement of the train, the overall amount of displacement of sand caused by the falling beanbag is increased. The beanbag literally contained more energy to be transferred to the sand on impact, creating a greater displacement of sand.

This increase of energy is an actual increase of energy that is added within the very structures of the atoms that make up the beanbag. Any change in momentum is a change in the overall energy level of a given mass. This means that the same beanbag in different inertial frames has differing amounts of energy.

So how can two different observers, such as observer A and observer B, measure the same falling object to travel two different distances? The difference in distances can be explained on the atomic level. The atoms of the beanbag dropped by person A had the same momentum pattern as the atoms of the moving train and person A on the moving train that dropped the beanbag. The atoms of person B, who was standing on the earth watching the train go by, had the same momentum pattern as the atoms of the earth upon which person B was standing. The atoms of the beanbag, person A, and the train were at a higher energy level than the atoms of person B and the earth. Person A observed the beanbag from the same energy level as the beanbag, while person B observed the beanbag from a lower energy level than the beanbag. Being at the same energy level as the beanbag, person A sees the beanbag fall straight down as observed in Galilean relativity. Being at a lower energy level than the beanbag, person B sees the beanbag fall at an angle towards the ground. From their immediate energy level in relation to the energy level of the beanbag, the measurement of both observers is absolutely correct even through the beanbag took only one absolute path through space.

When the beanbag was dropped from the moving train, the atoms of the beanbag maintained the momentum pattern of the moving train until impact. Upon impact, each atom of the beanbag experienced a quantum adjustment as they switched inertial frames. In this case, the momentum pattern of each atom of the beanbag lost energy as they adjusted to become synchronized to the new momentum pattern of the atoms of the earth.

In summary, the faster a beanbag (or any mass) moves relative to the earth's inertial frame, the more energy it acquires. The energy level of mass changes when the inertial frame that carries it changes. As a train (or any mode of transportation carrying mass) accelerates to go from one inertial frame to the next, say from 5 mph to 10 mph, its mass (and any mass that appears to be resting on it) literally acquires more energy within its atomic structures. This causes observers of differing energy levels to observe the motion of the same mass differently. For example, observer B, who was at a lower energy level than the beanbag, experienced the momentum of the beanbag through space differently then Observer A, who was at the same energy level as the beanbag. The differing energy levels of the observers account for the differing distances of travel by the same mass.

In Galilean relativity, from a macrophysics perspective, there appears to be no differences in the laws of physics for the same mass in differing inertial frames. A ball always falls directly towards the earth's surface in all inertial frames. From a microphysics perspective, there are differences in the energy level of the same mass in different inertial frames. The same ball has differing energy levels in different inertial frames. Mass does not just mysteriously move through space. The science of motion accounts for all changes in speed and direction.

Quantum Relativity

How did the same falling beanbag travel two different distances? The answer is quite simple. The differences in distance had little to do with the beanbag, which traveled one absolute path through space, and had everything to do with the energy level (inertial frame) of each observer. Think of the following experiment that could be done on the space shuttle freefalling around the earth. Person A is floating from one side of the space shuttle to the other side. When he reaches person B, who is stationed in the middle, he lets go of a quarter. Because the individual atoms that make up person A and the quarter share the same momentum pattern and energy level, even after person A let's go of the quarter, person A and the quarter continue to move through space at the same speed and in the same direction. Person B, who is stationed in the middle, watches as the quarter slowly moves away at the same speed that person A is moving away. Person B sees the quarter travel a greater distance from his immediate perspective than does person A, who sees the quarter travel with him as if he were still holding it. Even though the quarter travels only one distinct path through space, each observer measures a different length of travel of the quarter from his or her immediate perspective. This is because Person A and the quarter remained at the same energy level, while person B observed from a different energy level.

Galilean relativity is nothing more than differences in the energy level of the atoms of the same mass in differing inertial frames. If Galileo could have measured the energy level of the atoms of any object in his boat cabin while his boat was tied to the dock, he could have compared that measurement to the energy level of the atoms of the same object when his boat was sailing. The measured difference would directly correlate to differences in speed for the same object in different inertial frames because the atoms of the same mass change their energy level when they go from one inertial frame to another inertial frame.

Quantum relativity operates on two principles:

1. The acceleration and deceleration of mass are only temporary stages between the equilibrium of uniform motion.
2. The acceleration or deceleration of mass is always accompanied by the absorption or emission of energy to exactly accommodate changes in the momentum pattern of each atom of that mass. This allows each atom to adjust to the new energy level required to maintain its new momentum pattern.

Quantum relativity is an integral part of the science of motion. The differences in the energy level of the same object in different inertial frames are as much a measured science as differences produced by chemical reactions. As with any change in the physical properties of any mass, nothing is added to or taken away without a full accounting and explanation.

The reason the laws of physics are the same for all inertial frames is because the atoms of all the masses of the same inertial frame share the same momentum pattern and energy level. Just as the atoms of a mass are synchronized in their momentums, the masses in the same inertial frame are also synchronized in their momentums.

The Problem with Special Relativity

Quantum relativity would be incomplete without explaining the problem with Einstein's Special Relativity. Einstein's dilemma was similar to the beanbag scenario. Replace the beanbag with a light pulse. Person A, on the moving train, sees a light pulse go four feet. Person B, standing on the stationary earth relative to the moving train, sees the same light pulse travel more than the four feet. This is because the light pulse will not only travel the four feet to the same target, but will also travel at an angle proportionate to the speed of the train. Steven Hawking put it like this:

> Since the speed of the light is just the distance it has traveled divided by the time it has taken, different observers would measure different speeds for the light. In relativity, on the other hand, all observers *must* agree on how fast light travels. They still, however, do not agree on the distance the light has traveled, so they must therefore now also disagree over the time it has taken. (The time taken is the distance the light has traveled – which the observers do not agree on – divided by the light's speed – which they do agree on.) In other words, the theory of relativity put an end to the idea of absolute time! (Hawking, 1996, 21-22.)

How does light relativity compare to Galilean (beanbag) relativity? First of all, both the light pulse and beanbag were measured to travel different distances by different observers, so this is the same. Second, both observers agree on the light's speed (since supposedly this is constant for all inertial frames), but they disagree on the beanbag's speed since the beanbag traveled a further distance according to observer B within the same amount of drop time, so this is not the same. And third, because the time taken for the light pulse is the distance traveled divided by the light's speed, each observer experiences the same light pulse in a different amount of time, whereas both observers of the beanbag measure the beanbag traveling from start to finish in the same amount of time, so this is different, too. Both differences (speed and time) are the result of the mathematical assumption that the speed of the light pulse is the same for both observers, yielding different times for the same event. Remember, you can prove anything with mathematics. So why did Einstein make the assumption that the speed of light would be the same for both observers in this scenario?

Einstein built his Special Theory of Relativity on two postulates. His first postulate states that the laws of physics are the same for all inertial frames. This correlates with Galilean relativity wherein a person in a closed cabin of a ship cannot tell his relative motion to the land by appealing to the macro laws of physics. Einstein follows that up with his second postulate by stating that the speed of light is the same for all moving objects. This correlates to James Clerk Maxwell's mathematical discovery that electromagnetic waves travel at a fixed, unchanging speed. (Remember, with no ether, what was light's fixed, unchanging speed relative to?) When Einstein combined the two postulates, he theorized that the speed of light was constant, meaning the same, for all inertial frames. In other words, the speed of light was the same for all moving objects—a walking person, a flying airplane, a speeding rocket, and an orbiting earth. Einstein was able to accomplish this monumental task by inserting time into the equation. The faster an object moved through space, the slower its time frame. Time slowed for moving objects in direct proportion to light passing it so that all moving objects experienced the speed of light at its constant speed as calculated by Maxwell. Since this elaborate theory was formulated under the spell of the motion myth paradigm, it completely bypasses the role of quantum physics in the motion of mass. For this reason, it contains some false assumptions.

To answer the dilemma proposed by Stephen Hawking, I ask the following question: *What is the major difference between a light pulse and a beanbag?* The beanbag can appear to be at rest in multiple inertial frames from a macrophysics perspective. On the other hand, an emitted light pulse is never at rest in any inertial frame. This means that an emitted light pulse operates differently than mass, such as a beanbag. For example, it is common knowledge that the speed of light is unaffected by the speed of the mass emitting it, but the speed of an object is directly affected by the speed of the mass from which it is released. This major difference is due to the fact that in Galilean and Newtonian physics, objects can acquire a perceived state of rest within an inertial frame—meaning synchronized momentum patterns—whereas a freely moving light pulse never finds rest within any inertial frame.

Let's quickly diagram the problem.

Illustration--5
Light Clocks

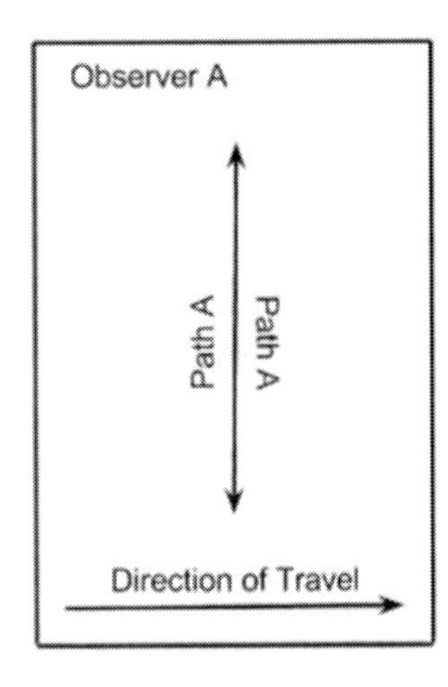

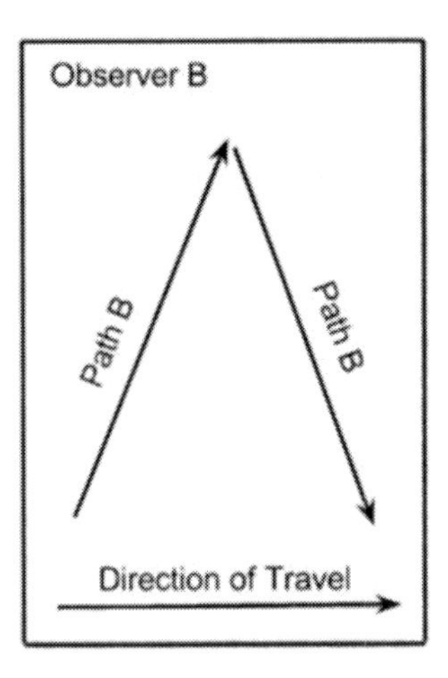

If *observer A* is traveling in a train with a hypothetical light clock, the light will move in an up and down motion in reference to *observer A* (Path A). O*bserver B*, who is stationary to the motion of the train that is carrying *observer A* and his light clock, will observe the path of the light moving at angles in the direction of the uniform motion of *observer A* and his light clock (Path B). The distance of *path A* is different than the distance of *Path B*. Yet, from each person's perspective, they are both correct. This is quite a paradox if the speed of light is constant for all observers and yet travels two different distances. Einstein's idea was to take the different distances of travel and divide them by the same light speed; you end up with two different times for the same event. Mathematically, it makes sense.

As pointed out earlier when different observers watched the same beanbag fall, the speed of the beanbag was different for each observer because each observer observed from a different energy level. This accounted for the different distances that the same beanbag traveled as measured by each observer even though the beanbag only traveled one distinct path through space. For light relativity, Einstein's epiphany was that each observer experienced the same event from different time frames rather than different energy levels.

It is important to emphasize that in the beanbag scenario, the differences in distance are due to differences in the energy level of each observer in relation to the energy level of the falling beanbag. This makes complete, logical sense from a quantum perspective. In Special Relativity, Einstein inserts time to replace energy levels. Remember, Einstein had no concept of quantum processes governing changes in momentum. He was stuck in a motion myth paradigm mode. Mass just mysteriously went from one inertial frame to another inertial frame with no accountable science. There was no measurable difference for the energy level of the same object in different inertial frames. For this reason, he gave light the same Galilean relativistic qualities as mass. In other words, he assumed that light would follow the same path as mass when in uniform motion. This led to the same light pulse traveling different distances for each observer. Then he applied his postulate that the speed of light is the same for all observers. For Einstein, because *Observer A* and *Observer B* experience the speed of light at its constant speed, they each experience *time* differently in order to compensate for the different distances the same light pulse traveled. Einstein mathematics validated his point.

Illustration—5 is a classic example used to explain time dilation of Special Relativity in many up-to-date encyclopedias. In illustration—5, if you replaced the light pulse with a bouncing ball, this would be an example of classic Galilean relativity for which I have already provided an explanation on a quantum level. However, a light pulse and a bouncing ball are not the same, so you shouldn't expect the same results. Einstein's mistake was to assume that a light pulse operated by the same rules as mass in Galilean relativity.

A friend and I spent many hours discussing this dilemma: Does light have the same relativistic qualities as a bouncing ball? Knowing that the speed of light has been proven to be independent of the speed of an object emitting it, (Alväger et al. 1964), we came to the conclusion that a light pulse's direction must also travel independently of the uniform motion of the mass from which it is emitted, realizing this consequential argument:

If the speed of light is independent of the motion of the mass emitting it,
then its direction or path should also be independent of that motion.

This answered the light pulse dilemma. The speed and direction of emitted light pulses are unaffected by the motion or momentum of the objects emitting them. An object of mass, such as the beanbag, which can be at the same momentum level as the observed inertial frame within which it is contained, must be treated differently than freely moving light pulses, which are never at the same momentum level of any observed inertial frame. In other words, freely moving light pulses cannot be regarded in the same manner as mass in respect to Galilean relativity and inertial frames.

Illustration--6

Different Views for the Direction of a Light Pulse
Emitted from Mass in Motion

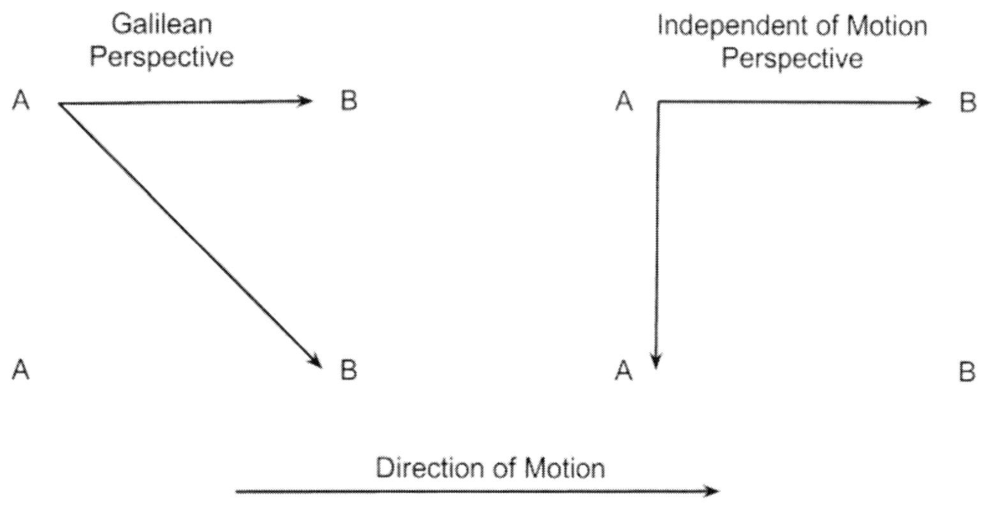

Galilean
Perspective

Independent of Motion
Perspective

A B A B

A B A B

Direction of Motion

In illustration—6, *A* represents uniformly moving mass in the direction of *B; A* also represents the emission of a light pulse perpendicular to the motion of the mass emitting it. The *Galilean Perspective* demonstrates light's direction or path of travel *dependent* on the motion of the mass emitting it—a continuation of Galilean relativity. The *Independent of Motion Perspective* demonstrates light's direction or path of travel independent of the motion of the mass emitting it.

Just as the speed of light is unaffected by the motion of the mass emitting it, the *path or direction* it travels after its emission should also be unaffected by the motion of the mass emitting it. This is because light does not have relativistic qualities, whether confined within mass or freely flowing in space. It goes back to comparing the motion of a photon to the motion of an atom. A photon has one fixed speed through space as calculated by Maxwell, whereas an atom's speed through space varies according to the momentum pattern and energy level of the protons and neutrons of that atom. For this reason, when mass is pushed away from a moving body, its new momentum is added or subtracted from its old one like throwing a baseball from a moving vehicle. On the contrary, the speed of light is unaffected by the speed of the object emitting it, such as turning a flashlight on from the same moving vehicle.

I remember someone telling me how Special Relativity was demonstrated in a college class. One student took a piece of chalk and perpetually drew a line going up and down while staying in one place. The second person did the same up and down motions as the first person while walking along the chalkboard from one end to the other end. This supposedly demonstrated how light traveled different distances in different inertial frames. Unfortunately, this isn't how light works. The speed of light is not the same for differing inertial frames as postulated by Einstein, (who assumed light had relativistic qualities similar to mass), but rather, the speed of light operates independently of all inertial frames. Once light is emitted from the confined energies of inertial mass, its *speed* and *direction* of travel move independently from the momentum of the mass emitting it.

Of all people, why would Einstein make this crucial mistake, assuming the speed of light is the same for all inertial frames? He was stuck in a motion myth paradigm. For Einstein, because objects just mysteriously move through space, Galilean relativity remained a mystery. No one, including Einstein, could explain why the laws of physics are the same for all inertial frames. From this premise, Einstein assumed that if the laws of physics are the same for all inertial frames, then the fixed, unchanging speed of light as calculated by Maxwell must also be the same for all inertial frames. It was a terrible miscalculation on Einstein's part that reflected the scourge of the motion myth paradigm.

When Galilean relativity is explained from a quantum perspective using momentum patterns and energy levels, then one can easily see that a photon operates differently than an atom. An atom can obtain an apparent state of rest in differing inertial frames, whereas a photon never obtains an apparent state of rest in any inertial frame.

The Independent Nature Light

When I was first developing the quantum model of motion, I thought the independent nature of light in and of itself might account for the null results of the Michelson and Morley experiment. I focused on the independent nature of light moving through the apparatus while completely ignoring what *effect* the independent nature of light confined within mass could have on the structure of mass as it moves through space. Like the rest of the science community, I took for granted the unenlightened view that mass mysteriously moves through space. I had no idea that the confined energies that make up mass could also be the cause and source of its momentum through space.

When I learned that the independent nature of light alone did not solve the Michelson and Morley dilemma, I started wondering if the independent nature of light confined in mass might be the cause for the contraction of mass calculated by Fitzgerald and Lorentz. Then it hit me. *What if the energies that make up mass are responsible for its motion through space, accounting for the contraction of mass at differing speeds?* I then wondered if free-flowing energies confined by the strong force within the atomic structure of mass provided the sensitivity to initiate quantum changes due to changes in momentum. This is when the quantum model of motion really started to take shape. Free-flowing energy confined as mass could maintain sensitivity to the slightest changes in momentum and could adjust at the exact instant momentum changes begin to occur.

I then realized that Einstein was right about one of his postulates: The speed of light is the same for all moving objects. There is consistency in the universe. My explanation as to how this is possible differs greatly from Einstein's macrophysics perspective. His mistake was to view Maxwell's constant for the speed of light to be the same for differing inertial frames, viewing the motion of mass as a whole and not from its parts, a paradigm cemented by Newton. Einstein said that the same light pulse would pass a moving car or speeding rocket at the speed of light as if they were not even moving. He justified this by calculating that the motion of any macro object or inertial frame caused a distortion is space-time to the degree that light passing it maintained its fixed, unchanging speed relative to that moving object or inertial frame. According to Einstein, every moving objcct contained its own time frame in reference to the speed of light so that Maxwell's equations were not nullified. In this way, the speed of light remained constant for all moving objects and in all inertial frames. This definitely seemed like a bold, brilliant move.

How does the quantum model of motion address this very idea that the speed of light is the same for all moving objects, maintaining Maxwell's constant speed for all electromagnetic waves? First of all, if the speed and direction of light are independent of the speed and direction of any mass emitting it, then the speed of all free-flowing electromagnetic waves in space inherently move through space at the same fixed speed, notwithstanding its source of emission. This means that the speed of electromagnetic waves in open space is comparable with itself. All electromagnetic waves move through open space at the same fixed speed, the speed of light.

Secondly, when comparing the speed of light to moving bodies, do you compare it with the mass as a whole or with the energies that make up mass? Einstein compared it to the speed of objects as a whole. The quantum model of motion compares the speed of light to the energies that make up mass. The atoms of mass are made from energies that account for both their structure and momentum through space. These confined energies retain their independent, fixed speed through space even when the mass goes from one inertial frame to another. This is why quantum adjustments are needed when changing inertial frames. The energies don't speed up or slow down, they stay the same fixed, unchanging speed. These energies are organized as the structure and momentum of an atom through space. An atom can be sped up with a restructuring of that organization as energy is absorbed or slowed down with a similar restructuring of that organization as energy is emitted. In other words, even though an atom as a whole can regulate its combined, linear momentum through space by adding or subtracting energy in order to speed up or slow down, the energy that makes up an atom maintains its fixed, unchanging speed through space. So when light passes any macro object, it is already traveling at the same speed as the energy from which the macro object is made, notwithstanding the speed of that object through space.

Instead of comparing the speed of light to the motion of a macro object as a whole, as Einstein did, inventing space-time as a variable to keep the speed of light constant in relation to the motion of that macro object, the speed of light moving freely through space is already the same as the speed of the energy making up any mass. Instead of comparing a photon passing a moving rocket, the quantum model of motion compares the photon passing the energy that make up the mass of the moving rocket. This maintains the constant velocity of light throughout the universe, whether freely moving through space or confined as energy in mass. It is comparing light to the energy that constitutes mass rather than comparing it to our macro perspective of that mass. In summary, the unchanging, fixed speed of light is the same whether it is freely moving through the universe or confined as the structure of an atom or particle. It is light's fixed speed and independent nature that drives the atomic and quantum models of motion.

What is Energy?

Quantum relativity helps us understand how energy moves through objects. As I try to explain energy to my sixth grade students by making waves with a rope, I ask them: *What is it that is traveling through the rope?* Someone will usually give the answer of energy. *Then I ask: What is energy? How does energy make the rope move?* This question of how energy moves through the rope as it is whipped to make wave motions perplexed me. A reaction takes place, but why, or even more important, how?

An in-depth look into quantum relativity explains the answer to this question. The observed movement of the rope is caused by a chain reaction on the atomic level. Before the rope is whipped, the momentum patterns of the atoms from which the rope is made are in par with the momentum patterns of the atoms of an orbiting earth through space. As the rope is whipped, the momentum patterns of the atoms making up that part of the rope change. The momentum patterns of those atoms react immediately to this disruption by absorbing or emitting energy to exactly correspond to the change. This disruption of momentum patterns disrupts the momentum patterns of corresponding atoms along the length of the rope. You end up with atoms in the same rope at varying momentum patterns and energy levels. In other words, atoms within the same rope temporarily experience different momentum patterns as they absorb and emit energy in an attempt to reestablish and maintain equilibrium.

The chain reaction of changing momentum patterns of the atoms making up the rope creates the resulting and observable movement of the rope in the form of a wave. Because the atoms of the rope are bonded together, the disruption of the momentum pattern of one atom causes a disruption to another and another, creating the chain reaction throughout the entire length of the rope. What is happening at the speed of light on the quantum level, the absorption and emission of energy accommodating the momentum changes of atoms, we see as a comparatively very slow moving wave through the entire length of the rope. Quantum gravity quickly returns the rope to its original state when the source of the wavelike motions cease.

When you drop a stone in water, it appears that energy moves through the water in the form of waves. What is really happening is the absorbed energy that caused the stone to accelerate disrupted the momentum patterns of the lower energy level atoms of the water. The waves in the water are the chain reactions of disrupted momentum patterns and their subsequent absorptions and emissions of energy to compensate for the changes. The waves are the result of changes in the momentum of the atoms that make up the water molecules. The energy didn't need water to travel through. The energy was the water at varying different energy levels. This created the waves.

In summary, had Galileo been able to observe the energy levels of atoms, he would have noticed that they change to correspond with differing inertial frames.

Chapter 13

Quantum Gravity

Absorbed Energy Accelerating Mass

Gravity is typically explained as an attraction between masses. Quantum gravity explains this attraction as the emission and absorption of energy between the masses. The absorbed energy accelerates the momentum of atoms. An acceleration of atoms is an acceleration of mass.

Newton's mistake has proven fatal for many centuries in understanding how gravity really works. As a result of Newton's mistake, the motion myth paradigm has tightly gripped the minds of most, if not all, physicists. And, to this day, physicists cannot explain how gravity really works. That is until now.

Einstein's happiest thought was the realization that inertial acceleration and gravitational acceleration are equivalent. In other words, gravity is the acceleration of mass. He used an example of an elevator being pulled up in empty space at 32 feet per second per second to illustrate this point. Thirty-two feet per second per second means that for every second of acceleration, an object travels an additional 32 feet of the distance of the previous second. At the end of the first second, an object travels 32 feet. At the end of the second second, the object travels 64 feet within that second. At the end of the third second, the object travels 96 feet within that second and so on. The elevator being pulled in empty space would create the same gravitational acceleration that a person experiences standing on the earth or an object falling towards the earth. The reality that gravity is the acceleration of mass creates the basis for quantum gravity. (See Appendix 1: Einstein's Elevator.)

Quantum gravity is the fourth leg to the Quantum Model of Motion. Quantum gravity is the absorption of energy within atoms, causing acceleration of those atoms in the direction of absorption. It is an exact cause and effect science with energy absorptions correlating exactly with acceleration increases.

When Newton said that an object in motion tends to stay in motion unless acted upon by a force, he was obviously referring to an external force such as a collision or friction. The result of the external force is a change in the momentum of that mass. From a quantum perspective, an external force causes a change in the momentum pattern of each proton and neutron of an atom, which initiates a simultaneous change in the energy level of each proton and neutron of that atom. This happens every time an external force acts upon an object in motion. Energy is absorbed or emitted to exactly correlate with the momentum change.

The same effect occurs when energy is absorbed into the nucleus of an atom. A change in the energy level of each proton and neutron of an atom initiates a change in the momentum pattern. [Say the horses are spooked and they start running faster, (*energy level*), the overall speed of the yoked horses increases, (*momentum pattern*).] The result is the acceleration of that atom in the direction of absorption. When energy is added to the nucleus of an atom, its momentum simultaneously shifts to accommodate the increase in energy. This momentum shift is the acceleration.

Gravitational acceleration is equivalent to inertial acceleration. Both result in the change of the energy level and momentum pattern for each proton and neutron of an atom. An external force causes inertial acceleration, whereas the absorption of energy causes gravitational acceleration. Whether by an external force or the absorption of energy, the result is the same. The atom or mass accelerates. This is because a change in the momentum pattern (external force) changes the energy level of each proton and neutron, and vise versa, a change in the energy level (absorption of energy) changes the momentum pattern of each proton and neutron. Either way, the result is acceleration. The simultaneous effect of the momentum pattern changing the energy level or the energy level changing the momentum pattern is like blowing air into a balloon. As the balloon receives more air, the boundaries of the balloon expand. As the balloon loses air, its boundaries contract. (As pointed out in the quantum adjustment chapter, an external force can accelerate or decelerate the linear speed of an atom. Energy absorption always accelerates the linear speed of an atom.)

Since mass is nothing more than a collection of bonded atoms synchronized in their quantum momentums, an acceleration of atoms is an acceleration of mass. The constant flow of energy into the atoms of a mass creates an acceleration of 32 feet per second per second near the surface of the earth in the direction from which the energy is absorbed. The effects of the absorbed energy can be seen when an accelerating object decelerates at the moment of impact with a lower energy level surface such as the earth's surface. In the process of reestablishing equilibrium with atoms of a lower energy level, quantum processes transpire. This sets off a chain of events that can be witnessed at the moment of impact as will be explained in the falling penny example.

Gravity, as absorbed energy, changes the momentum pattern and energy level of each proton and neutron of each atom of a mass, accelerating the <u>already moving mass</u> in the direction of absorption.

This is why Newton's mistake was so crucial. If mass is perceived to be at rest when it is sitting on the earth like a car in the parking lot at a grocery store, then the atoms making up that mass are not perceived to be the cause of that mass's motion through space. This misperception blocks the potential for understanding how gravity really works. By correcting Newton's mistake, we now understand that the atoms that make up the mass of the car that appears to be resting on the earth's surface are in reality the cause and continuation of that car's motion through space. The atoms making up the mass of that car are in perpetual motion. What appears to be resting is actually mass in motion wanting to accelerate but experiencing terminal velocity due to the resistance of the earth's surface. If we could shut off the effects of gravity—the energy causing acceleration—the car would no longer be accelerating into the earth, nor would it be attached to the earth. A good nudge would cause the car to go off in a different direction than the earth, accentuating the motion it already has, much like the slightest nudge to the water container in the free-falling space shuttle changes its momentum until acted upon by another force.

The Falling Penny

In this example and throughout the rest of this chapter, when I refer to the momentum pattern of an atom, I am are referring to the momentum pattern of each proton and neutron of that atom. The reason I do this is to avoid the repetition of saying "the momentum pattern of each proton and neutron of that atom." An atom's momentum pattern refers to the momentum pattern of each proton and neutron of that atom.

An example of the effects of quantum gravity is dropping a penny to the floor. First of all, as you hold the penny above the floor, the penny is already in motion. It is sharing the same motion as the hand holding it, just as the quarter shared the same motion as the moving astronaut, even after he let it go. When you let go of the penny, the protons and neutrons of the atoms making up the penny absorb energy emanating from the surface of the earth. The momentum pattern of the protons and neutrons of each atom simultaneously change, increasing the overall motion of the penny in the direction of absorption 32 feet per second per second until it hits the ground.

When the penny hits the ground, its increased energy level from the absorbed energy that accelerated its momentum interacts with the lower energy level of the ground. At the very moment of impact, the momentum pattern of each atom changes. Energy is emitted from each proton and neutron of each atom to accommodate their new momentum pattern. At the initial bounce, the penny is off in a different direction. If not for gravity, the penny would appear to float off in its new direction until a force acted upon it. (To help visualize this, think of an object in the space shuttle that is floating horizontally towards one of its walls. When the object hits the wall, it changes its direction, and then continues in its new direction until a force acts upon it.) As the penny starts to go off in a different direction, it begins to absorb more energy emanating from the earth's surface, changing its direction and accelerating it towards the earth again to repeat the process over and over until the penny finally rests on the earth's surface, with its atoms sharing the same momentum pattern as the atoms making up the earth. During this whole process, the sum of mass-energy in the universe remains unchanged.

When the penny makes contact with the lower energy level surface, there is an immediate change in the energy level of each proton and neutron of every atom of the penny. This chain reaction carries into the atoms of the surface that were disrupted at impact, causing the absorption and emission of energy according to their equilibrium needs. (This is like the sand that was displaced in the sandbox when the beanbag landed in it.) When the penny finally appears to rest on the earth's surface, the momentum pattern and energy level of the atoms of the penny remain synchronized with the momentum pattern and energy level of the atoms of the earth until another disruption occurs such as when somebody picks up the penny.

The penny appears to stay at rest on the earth's surface because it is continuously absorbing and emitting energy at an even rate due to the terminal velocity reached because of its continuous contact with the surface of the earth. The energy necessary for acceleration is continuously being absorbed into the penny, but because the momentum pattern of the atoms cannot change due to their terminal velocity, the absorbed energy is emitted at the same rate of absorption.

Drop a penny a few times on the surface of a table. Watch at impact as the penny goes from a higher energy level (the energy acquired to accelerate it towards the surface of the table) to the lower energy level of the surface of the table, (the previous energy level of the penny before you picked it up and dropped it). Drop the penny from various heights to compare the amount of energy it acquires from the varying heights.

Quantum Gravity

Gravity is not a mysterious force of attraction between masses, nor is it the effects of warped space-time, but rather, it is energy (forces) acting within matter. It is the effects of absorbed energy on already moving mass. This is why exposing the motion myth and Newton's mistake is so important to understanding how gravity really works. Without this step, you cannot understand how gravity is the absorption of energy accelerating mass that already has momentum.

This is why Einstein was unable to unify general relativity (his explanation of gravity) with quantum physics. Like Newton, he failed to recognize that all objects are always in motion, the first major step for understanding how gravity really works. Instead, he brilliantly created space-time as a substitute for quantum processes, accurately predicting the effects of gravity without addressing the quantum causes.

Space-time describes distortions in space caused by large masses such as suns and planets, and the effect this has on moving objects near their surfaces, such as other planets, moons, asteroids, and light. In reality, these distortions in space are actually higher concentrations of energy near the surfaces of large spherical masses such as the sun or earth. This will be explained in more detail in the next chapter, which is titled The Energy Cycle.

Quantum gravity is only one part of a four-legged stool. The other three descriptive parts already discussed are quantum momentum, quantum adjustments, and quantum relativity. These four parts make up the Quantum Model of Motion. The Quantum Model of Motion explains why objects move in the first place. Without understanding why objects move in the first place, one cannot understand how mass is accelerated by the absorption of energy. *Quantum gravity can now be easily explained as the acceleration of mass caused by the absorption of energy. The absorbed energy changes the momentum pattern and energy level of each proton and neutron of each atom, causing acceleration towards the source of the energy being absorbed.*

Because the momentum of each proton and neutron is independent of the momentum of all other protons and neutrons, (like each individual horse yoked together with the other horses), the number of protons and neutrons in an atom doesn't change that atom's acceleration rate in comparison to atoms of differing numbers of protons and neutrons. All protons and neutrons of a mass are exposed to the same energy source and absorb the same amount of energy, causing the same momentum shift in the direction of absorption. This is why Galileo could drop two differing size balls off the same tower and watch them land at the same time. Because the larger ball has more protons and neutrons, its overall energy or inertia will be greater than the ball with fewer protons and neutrons. But the acceleration of both balls will be the same because all protons and neutrons accelerate at the same rate, for all protons and neutrons are exposed to the same energy concentration emanating from the earth. For this reason, all atoms, elements, compounds, and masses accelerate at the same rate. This demystifies the equal attraction law of gravity.

The Visible Effects of Quantum Gravity

The absorbed energy due to the effects of gravity is visible. When I hold a pencil several feet above a hard tile floor and drop it, the absorbed energy responsible for accelerating the atoms of the pencil causes the pencil to bounce a few times when it hits the floor. Energy is emitted from the atoms of the pencil with each bounce until the atoms of the pencil share the same energy level as the atoms of the surface of the floor upon which it now appears to be resting. When I held the pencil above the floor, it is said to have potential energy. Potential energy is nothing more than potential changes in momentum patterns—acceleration—before a terminal velocity is reestablished with the surface of the earth.

When a ball bounces off the ground, it moves away from the surface of the earth. It would continue in this direction uninterrupted if it weren't for energy being absorbed back into the atoms of the ball due to the effects of quantum gravity. After the ball reaches its apex, briefly sharing the same momentum pattern and energy level as the atoms of the earth, it begins to accelerate towards the earth again. With each bounce, more energy is emitted until eventually the ball comes to a resting position on the earth. You can visually see the effects of acquired energy causing the acceleration that precedes each bounce. When the ball appears to be resting on the earth's surface, it is still moving through space. The momentum pattern and energy level of each atom of the ball are still the source and cause of its movement through space. If the energy emanating from the earth shut off and I threw the ball into the air, it would move away from the earth like a helium balloon escaping from the grasp of a child.

When my car comes to a quick stop, the nametag hanging from my mirror continues to swing back and forth until the energy level of its atoms reaches equilibrium with the new energy level of the atoms of the car of which it is attached to by a string. Here is an interesting point about the swinging nametag. As the nametag's momentum causes it to swing towards the apex of its swing, it slowly comes to a stop. Energy has been emitted so that it temporarily shares the same momentum pattern as the car. At that moment, the effects of gravity accelerate the nametag towards the earth causing it to do a grandfather swing in the opposite direction to its new apex. Gravity then accelerates it again and the process continues back and forth until eventually it evens out and hangs straight down, sharing the same energy level as the car.

As absorbed energy accelerates mass towards the surface of the earth or as emitted energy accompanies the deceleration of mass when it appears to come to an abrupt stop, the effects and results are observable and ultimately measurable.

The Simultaneous Influence of Quantum Momentum and Quantum Gravity

How do objects bound by gravity to the earth's surface easily move in any direction when a force is applied? Think of a marble resting on a floor. If you apply the same force in any direction, it will go the same distance. The atoms of the marble resting on the floor share the same momentum pattern and energy level as the atoms of the floor. They are moving at the same speed and in the same direction through space. If their momentums are synchronized, it seems strange that the marble can move freely in any direction when a force is applied. For me, understanding how a marble moves freely was very difficult to visualize. I had to temporarily forget about the earth and imagine different objects moving in the same direction in space, such as a peach and an apple moving side by side in the space shuttle. If you gave one of them a nudge in a perpendicular direction to movement, without interfering with its original momentum, the momentum pattern and energy level would adjust to its new directional change while still maintaining its original momentum. It would move away at an angle from the other fruit while still being lined up with it. In other words, they would cross the finish line at the same time. They would just be further apart. This same effect is like dropping luggage out of an airplane. As the luggage accelerates towards the earth, it maintains the original momentum of the airplane while simultaneously accelerating towards the earth.

When the marble appears to be resting on the floor, it shares the same momentum as the earth through space. Because the marble continues to accelerate into the floor, due to the effects of quantum gravity, its acceleration keeps it attached to the floor. This is like the apple that appears to be resting on the elevator floor when the elevator floor is accelerating into it. (See Appendix 1: Einstein's Elevator.) In this case, the marble is accelerating into the floor. This allows the marble to move in any direction on the floor with the same force because it is simultaneously experiencing quantum momentum and quantum gravity. Quantum momentum synchronizes the momentums of the marble and the floor. Quantum gravity accelerates the marble towards the floor. The marble is experiencing terminal velocity as it continuously tries to accelerate into the floor while sharing the same momentum through space as the floor. When the marble appears to be resting on the surface of the floor, the energy level and momentum pattern of each proton and neutron are at terminal velocity, evenly absorbing and emitting energy. The protons and neutrons are always juiced with the energy necessary to accelerate. Meanwhile, their energy level and momentum pattern are at equilibrium with the energy level and momentum pattern of the protons and neutrons that make up the mass of the floor. This means the marble is sharing the same uniform motion of the floor while juiced to accelerate into it. For this reason, you can apply the same force in any direction to the marble, and it will experience the same acceleration.

The second thing I did to help visualize this was to imagine a wall in space emitting gravitational energy. If you let the marble go several feet away from the wall, it would accelerate towards the wall. When it reached the wall, it might bounce a few times and then the marble would appear to rest against the wall. In actuality, the energy level and momentum pattern of each proton and neutron of the marble would be at a terminal velocity against the wall. The protons and neutrons of the marble would share the same momentum through space as the protons and neutrons that make up the wall. At the same time, the protons and neutrons of the marble are juiced to continuously accelerate into the wall. In this state, the same applied force would move the marble the same distance in any direction on the wall. It is like the apple and peach example a few paragraphs ago. When you nudge one of the fruits, it changes its direction slightly while keeping its original momentum. It wouldn't matter which direction you nudge it, it would still cross the finish line at the same time as the other fruit. In essence, the wall acts like the surface of the earth. If you flicked the marble with the same force in any direction, it would go the same distance. If you picked it up and let it go, it would accelerate towards the wall. When it hit the wall, it would bounce a few times, emitting energy until the energy level and momentum pattern of the atoms of the marble matched the energy level and momentum pattern of the atoms of the wall. Then it would again be in a state of terminal velocity against the surface of the wall, appearing to be at rest on the wall.

Tides

Another effect of gravity that is observable but difficult to explain is the cause of tides. Heretofore explained as the gravitational pull of the moon on the earth can now be explained as concentrated gravitational energy emitted from the moon and absorbed by the protons and neutrons of the atoms making up the water. This causes a shift in their momentum patterns, accelerating them in the direction from which the energy was absorbed until a terminal velocity is reached between the energy emitted by the moon in contrast to the energy emitted by the earth. The moon doesn't mysteriously attract the water nor does warped space cause the water to move towards the moon, but rather, it is energy emanating from the large, spherical-shaped moon accelerating the mass of the water towards the direction of absorption.

Satellites

Quantum gravity keeps satellites orbiting their source. Whether it is our solar system orbiting the center of our galaxy, the earth orbiting our sun, the moon or satellites orbiting the earth, quantum gravity keeps them in orbit. As the energy cycle keeps the flow of energy moving through spherical bodies, the energy emanating from their surfaces accelerates bodies in motion within a reasonable proximity of their surface in an orbital pattern. It is because these satellites already have motion that energy emanating from the surfaces of large spherical bodies can accelerate their motion into an orbital pattern. This is why exposing the motion myth and Newton's mistake is absolutely necessary in order to understand how gravity really works. If we do not understand the science of motion, we will never understand the cause of gravity.

Summary

Einstein's great epiphany that gravity is the acceleration of mass is a great starting point for understanding how gravity really works. But then one must also make the connection that all mass is always in motion. This was Einstein's roadblock. The motion myth paradigm, along with Newton's mistake, kept him from realizing the simple reality that all mass is always in motion and that gravity is just the acceleration of that motion.

Gravity is not a function of space-time, nor is it an invisible, unexplainable attraction of masses, but rather, gravity is the absorption of energy into the protons and neutrons of the atoms making up mass, initiating and sustaining the acceleration of already moving mass in the direction of the absorption.

This is how gravity really works.

[I am absolutely confident that all of Einstein's results-based mathematical accuracies predicting outcomes such as time dilation in Special and General Relativity theories will be able to be explained by quantum processes outlined in the quantum model of motion. Einstein's misuse of time as a substitute for quantum processes reflects the motion myth paradigm that mass just mysteriously moves through space.]

Chapter 14

The Energy Cycle

How does the earth or any sphere sustain the constant flow of energy emanating from its surface in order to maintain the constant effects of gravity? The earth is already in momentum around the sun. The amount of energy it receives from the sun allows it to continuously accelerate towards the sun without accelerating into it. As energy is absorbed into a spherical mass, mass either accelerates or—like the penny experiencing terminal velocity as it rests on the earth's surface—acceleration is restricted and energy is absorbed and emitted at the same rate. In spherical masses, the space necessary for atoms to change their momentum is limited. This means that the constant flow of energy being absorbed by spherical masses accelerate the atoms of the spherical mass until they experience terminal velocity as their momentums are restricted by the other atoms that make up the spherical mass. At terminal velocity, the atoms absorb and emit energy at the same rate. Eventually, the energy emitted by the limited momentum of the atoms that make up the spherical mass has nowhere to escape except through the surface of the spherical mass, and when it reaches the surface, it is then emitted into space. Emitted energy from any mass causes other masses around it to accelerate towards the first mass. This continuous process of mass emitting energy causing other masses to accelerate towards it produces the spherical shapes that we see in the form of stars and planets.

The absorbed energy that keeps the earth accelerating around the sun eventually has nowhere to go or escape except through the surface of the earth, moving out into space. This causes any object on or near the surface of the earth to accelerate towards its surface—gravity. The effects will be strongest closest to the surface and will gradually get weaker as energy spreads out the further it gets away from the spherical surface. Our earth and other planets in our solar system are beneficiaries of large amounts of this energy from the sun. Our sun receives this energy from all sources that absorb and emit such energy like stars in our galaxy, the center of our galaxy, and all the other galaxies in the universe. It is a reciprocal process that keeps energy moving throughout the universe.

Illustration--7

Energy Emitted from Spherical Surface Area

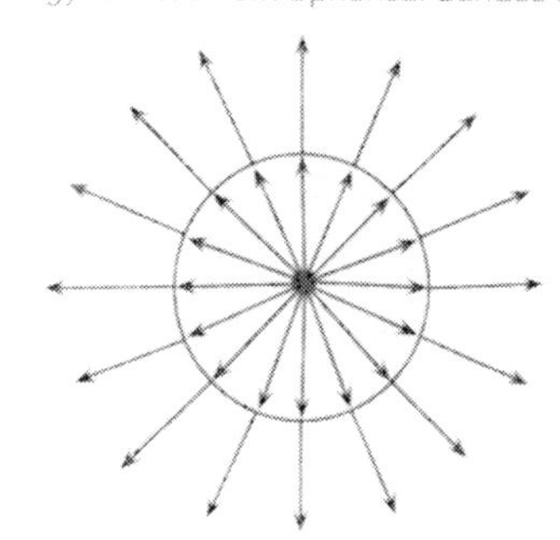

Due to terminal velocities restricting the acceleration of mass within a spherical body, absorbed energy is simultaneously emitted from the atoms of the mass comprising the spherical body until the energy finds its way to the surface of the spherical body and is emitted out into space. This creates concentrated amounts of energy near the surface of a spherical body. As this energy moves further away from the surface, its concentrated amounts are proportionately diluted. Through this cycle of energy, spherical bodies provide the source of energy that sustains quantum gravity near their surfaces.

Physicists before Planck's quanta theory of electromagnetic waves thought there was a substance that filled the immensity of space called ether through which light propagated. Interestingly enough, the universe is actually filled with energy moving at the speed of light, which is being absorbed and emitted through a reciprocal process; it doesn't need a medium to move through space as once theorized by the ether theory. Some areas of space have higher concentrations of this energy such as near the surfaces of large massive spherical bodies like the sun and earth. The high concentration of energy near the surfaces of spherical bodies explains Einstein's use of space-time to predict the movement of masses or the bending of traveling light near their surfaces. The distortions described by space-time are actually concentrated amounts of energy in comparison to the less concentrated amounts of energy in open space. The surface area of any spherical body will emanate the largest concentrations of energy before it gets diluted as it spread out in open space the further it gets from the surface.

Nature has a way of sustaining important process through cycles. For example, the water cycle keeps a continual flow of moisture over the land. *The energy cycle* keeps a constant flow of energy throughout the galaxy and beyond. As part of the energy cycle, energy continually flows into and out of our solar system, keeping the planets orbiting the sun and the moons orbiting the planets. It provides large concentrations of energy near the surfaces of large spherical bodies like our earth, allowing for a constant flow of energy emanating from their surfaces to sustain the perpetual acceleration of mass we call gravity.

Part IV

Conclusion

Chapter 15

Proof

Shifting Paradigms

Although many great scientific advancements have occurred since Einstein's great epiphany, we still cannot explain how gravity really works. Billions of dollars have been spent breaking matter up into smaller and smaller particles, and we still cannot explain how gravity really works. The Standard Model remains incomplete because it still fails to explain how gravity really works. Get the picture? After all these years, we still cannot explain how gravity really works.

Our inability to explain how gravity really works is based off a misperception that happened hundreds of years ago. Newton's mistake led to the motion myth, which led to Einstein's real blunder, which led to quantum mechanics being misguided, which led to the Standard Model being incomplete, which led to gravity still being unexplained. Taking for granted that objects move through space turned a blind eye on the need for a quantum model of motion. In order to move forward we must step backwards hundreds of years to correct Newton's mistake and the motion myth. Then we can apply what we have learned over the past several centuries in the light of these corrected misperceptions.

If the inherent logic of the quantum model of motion in not enough proof for its validity as a working model for the motion of mass through space, then I will offer further proof. The validity of the Atomic and Quantum Models of Motion hinges upon the starlight displaced by the gravity of the sun. Einstein pointed to the warping of space-time as the cause of this validated phenomenon, (Eddington, 1920). Quantum gravity, as explained in this book, theorizes that the bending of light around large massive objects like the sun is not caused by warped space, but rather, it is caused by large concentrations of energy emanating from the surface of spherical bodies like the sun. As light passes the surface of the sun, each photon absorbs some of this energy, causing a frequency shift, slightly changing their direction towards the source of the absorbed energy. The quantum model of motion theory theorizes that the displaced starlight will be blue shifted from the same starlight that isn't displaced, validating the underlying principle from which the Atomic and Quantum Models of Motion are built upon: Energy transfers accompany changes in speed or direction.

Frequency Shifts

This theory postulates that the frequencies regulating quanta can shift with an accompanying absorption or emission of energy. Any frequency shift must be accompanied by a proportional absorption or emission of energy to exactly compensate for the shift, or the conservation of mass-energy would be violated.

In Conclusion…

As the quantum model of motion continues to unfold, for this book is only the starting point, the more I appreciate Albert Einstein's Relativity theories. Einstein was one thought away from being able to correct his theories and explain them from a quantum perspective. Had he realized that mass is always in motion, even when it appears to be at rest, he would have broken the curse of Newton's mistake and the motion myth. He would have been forced to look at the quantum causes regulating momentum, relativity, and gravity. Although his brilliant theories mathematically predict the results of relativity and gravity, they perpetuate the quantum gap.

Einstein's failure to see that all mass is always in motion, the starting point for the quantum model of motion, prevented him from formulating a unified theory. He remained stuck in the motion myth paradigm. And now modern scientists are stuck in the same mindset and will be until they are able to break the curse of Newton's mistake and the motion myth. If this book does anything, it exposes Newton's mistake and the motion myth. It exposes the greatest misperception and stumbling block to understanding how gravity really works.

In the end, the purpose of science is not to compose reality to match our perceptions but rather to change our perceptions until they conform to reality—to see things as they are even if it is different from how we want them to be.

Chapter 16

Einstein Continued...

I dedicate the quantum model of motion to the continuation of Einstein's work that occupied the latter part of his life even to his dying day. Einstein intuitively knew that quantum physics alone could not resolve the quest for a grand unified theory. His quote about the moon still being there when he was not looking at it was a slam to the craziness of quantum physics. What Einstein didn't realize was the solution did not depend upon a mathematical theory or the discovery of a new particle. But rather, the continuation of his work depended on a paradigm shift in the way we view the motion of mass through space.

In my first book, *Einstein Continued...*, I referred to this new way of viewing the motion of mass as the missing model of motion. It implied that a model of motion was missing from the science of physics and was the key to connecting Einstein's relativities to quantum physics. It suggested that the model of the atom was incomplete without explaining how the energies that make up an atom also account for how that atom moves through space, and how those energies compensate for changes in speed and direction. Somehow, the science of motion was completely ignored and remained hidden in ignorance. The science community just falsely assumed that an object in motion stays in motion, for no apparent reason, until a force acts upon it.

It is sad to say that the mathematics of Einstein's own relativity theories buried the potential for discovering the science of motion. When Einstein referred to time or space-time in his theories, he was really referring to hidden quantum processes he couldn't explain at the time, quantum processes that could only be found in a quantum model of motion, the science of motion. If he were alive today, I am confident that he would have acknowledged his error and then continued his work in the light of the quantum model of motion to finish his work of explaining the universe in terms of a grand unified theory.

The insights Einstein needed to bridge the quantum gap, to translate space-time into quantum processes, is the same insights needed to understand how gravity really works. And to understand the motion of gravity, you have to understand how the science of motion is perpetually sustained by the energy cycle.

I dedicate the insights found in this book as a continuation of Einstein's work, which is a continuation of Newton's work.

Appendix 1

Einstein's Space Elevator

From a Quantum Model of Motion Perspective

Another way to explain quantum gravity is to think of Einstein's elevator example. Think of a person standing on an elevator floor in outer space, holding an apple. According to quantum momentum, the person, the apple, and the elevator floor are all moving through space at the same speed and in the same direction. (It might help to visualize this in a very, very large space shuttle freefalling through space around the earth.) If the person let go of the apple, it would stay in space right where he placed it because the atoms of the person and the apple have equivalent momentums through space like synchronized swimmers.

Now let's assume we start pulling up the elevator floor in the direction of the person and apple, 32 feet per second, per second. As the floor is being pulled up, its acceleration will run into the quantum momentum of the person and the apple. As the floor accelerates into the person holding the apple, it then accelerates the motion of the person and the apple with it. The constant acceleration of the floor simultaneously causes a constant acceleration of the person and the apple. The atoms of the person and the apple continuously try to reestablish the equilibrium of quantum momentum only to be accelerated again and again by the force of the accelerating floor.

Now, here is the interesting phenomenon. When the person lets go of the apple at shoulder height, the apple experiences a brief moment of no acceleration. At the exact moment the person lets go of the apple, there is no more force acting on it. It enjoys a moment of quantum momentum, moving through space with the exact momentum of when the person let go of the apple. This is like the astronaut setting the toothbrush in space next to him and it stays in that exact spot relative to the astronaut. What happens next is the floor continues to accelerate right into the apple so that the apple makes contact with the floor. Now the apple starts accelerating again with the floor, appearing to be attached to it. It would have the appearance of being at rest on the elevator floor. If you had a video of when the person let go of the apple, it would look like it fell to the floor, like the apple had a motion in a downward direction. In reality, the floor moved into the apple. At contact, their accelerations became synchronized again.

Once you stopped accelerating the elevator floor, quantum adjustments would stop, the equilibrium of quantum momentum reestablished, and the person, the apple, and the floor would continue to move through space, like synchronized swimmers, in the same direction at the same speed, with the atoms of each mass being the source and cause of their continual momentum through space.

Appendix 2

Gravitational Energy

Quantum gravity is not a blind theory wherein there is no physical evidence supporting its probable validity. As an object accelerates towards the surface of the earth, it acquires energy. We see the acceleration. At impact, there is a visible demonstration of the object's increased energy as it adjusts (usually making smaller and smaller bounces) until it reaches equilibrium with the lower energy level of the surface upon which it finally appears to rest. We literally see the effects with our eyes.

The question I am presently unsure about is the type of energy absorbed into an atom to cause gravitational acceleration? This is where my limits as a philosopher are exposed. I am not sure if the energy emanating from the earth, moon, and sun—causing gravitational acceleration—is within a detectable range with our most sensitive instruments, or is it a purer form of energy that is near or outside the detectable range of our most sensitive instruments? Whatever the refined nature of this energy may be, when it is absorbed it causes mass to accelerate. And when it is emitted from mass, other masses that absorb this energy accelerate towards the source of emission.

This book answers many questions that have perplexed physicists for many years. Unfortunately, it also leaves a few questions unanswered. As one mystery is unraveled, it reveals new mysteries, hidden within the unraveled mystery.

Bibliography

Barlow, Philip L. Spring 2007. Toward a Mormon Sense of Time. *Journal of Mormon History* Vol: 33, No. 1: 1-37.

Brian, Denis. *Einstein: A Life.* New York: John Wiley & Sons, Inc. 1996.

Folger, Tim. 2000. From Here to Eternity. *Discover* December: 54-61.

Hawking, Stephen W. *A Brief History of Time.* New York: Bantam Books. 1996.

Robinson, David and Groves, Judy. *Introducing Philosophy.* London: Icon Books. 1999.

Tolle, Eckhart. *The Power of Now.* Novato. 1999